EXPERIMENTAL
FLASH STEAM

EXPERIMENTAL
FLASH STEAM

J. H. BENSON
and
A. A. RAYMAN

△TEE△

TEE Publishing

WARNING

Since this book was first published, the Health and Safety at Work Act (1974) has led to much greater emphasis on safety in workshops which has extended to many leisure activities. Materials such as asbestos cannot be used for insulation but substitutes are available.

Speeds of model hydroplanes now exceed 100 m.p.h. resulting from much development work. The reader, in pursuing construction and operation of all flash steam plants, must exercise great care at all times and must accept that safety is your responsibility.

First published 1973
Reprinted 2000

ISBN No. 1 85761 116 0

CONTENTS

INTRODUCTION

INTRODUCTION

Many years have passed since the publication of the late Edgar T. Westbury's handbook *Flash Steam*. In view of present revival of interest it was thought appropriate to produce this new version.

The amount of information available to those who wish to know something about flash steam is very limited, unless much research is carried out. For example, *Model Engineer* has published many articles on the subject in a period covering well over 60 years, but few are fortunate enough to have access to complete volumes.

The subject matter is largely concerned with flash steam applied to model boats and hydroplanes seen from the viewpoint of the model engineer—as, indeed, was the earlier handbook. However, these miniature plants have similar problems to some of the larger ones and it is known that model designs have been scaled up and used in full-size applications.

We would hesitate to be dogmatic on this subject. We must emphasise that our opinions are by no means fixed or unalterable and we apologise in advance for errors and omissions.

A certain amount of matter has had to be left out. The omitted material is mostly of unproven character, and perhaps best avoided in a handbook.

Grateful thanks are offered to Model and Allied Publications Ltd., for their help, and in allowing us to use diagrams and material from past issues of *Model Engineer*, to The Institution of Mechanical Engineers for extracts and diagrams from the Proceedings of the Institute and also to the following who have all offered assistance with photographs and/or material: Prof. D. H. Chaddock, C.B.E., M.Sc., C.Eng., F.I.MechE., Thos. Hindle, Esq., Alec Hodsden, Esq., L.P. Purple, Esq., A. W. Cockman, Esq., J. Bamford, Esq., F. Jutton, Esq., G. Rosekilly, Esq., B. Squire, Esq., H. W. Saunders, Esq., and The British Light Steam Power Society.

Since writing the above, we are very sorry indeed to record the passing of Thos. Hindle and Bert Squire. Both gentlemen were life-long steam enthusiasts, and their expert advice and opinions will be greatly missed.

J. H. BENSON and A. A. RAYMAN

CHAPTER ONE

PRINCIPLES

Flash Steam! These magic words have caught the imagination of steam lovers and other enthusiasts for well over 60 years. It is perhaps not generally realised that the principle was probably first employed by Jacob Perkins early in the 19th century. This gentleman is best remembered for his invention of the steam gun. It was capable of discharging musket balls at 1,000 per minute, and the Duke of Wellington turned it down as being 'too destructive'!

The principle is very simple—a continuous length of tubing is heated, water forced in one end is 'flashed' into steam during its passage and emerges at the other end (Fig. 1.1).

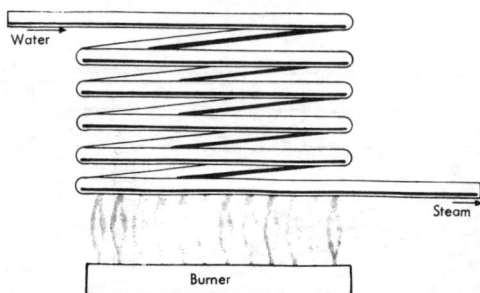

Fig. 1.1   A simple flash steam boiler

Since the earliest days of steam it has been known that the simple 'pot' boiler is only capable of producing a limited amount of steam in a given time. 'Forcing' by applying more heat results in violent ebullition causing severe priming without a commensurate increase of steam. 'Priming' means the carrying over of water with the steam, which not only reduces efficiency but can cause severe damage in some engines.

1

Increasing the overall size of such a boiler may present safety hazards since the stress on a circular drum increases with diameter, thus requiring an increase in plate thickness. Early boilers were often made of cast-iron or wrought-iron plates riveted together, so it is hardly surprising that quite a number of them exploded—usually because someone had tied down the safety valve or other malpractice. James Watt was a noted opponent of high pressures, 15 *psi* being regarded by him as quite high enough!

The problem of generating steam more rapidly was met by increasing heating surface by means of tubes, either by passing hot furnace gases through the water as in the locomotive type, or circulating water through the fire in the case of water tube boilers.

A replica of the *Rocket* made in 1929 for Henry Ford, on the occasion of the centenary of the Rainhill Trials, was stated to possess remarkable steaming properties. There is little doubt that the multiple fire tubes, passing through the water space, contributed in large measure to the success of both the original and the replica. Working pressure was 50 *psi* and the boiler shell was made of $\frac{1}{4}$-inch wrought-iron plates riveted together. Certain other types of fire-tube boilers were known as the 'Cornish' and 'Scotch' boilers, both of which employed large fire flues. Some had short 'Galloway' tubes passing water to improve circulation (this arrangement is similar in layout to the well-known centre-flue boiler used in models). By the end of the 19th century 'Scotch' marine boilers were 16 feet in diameter, the outer

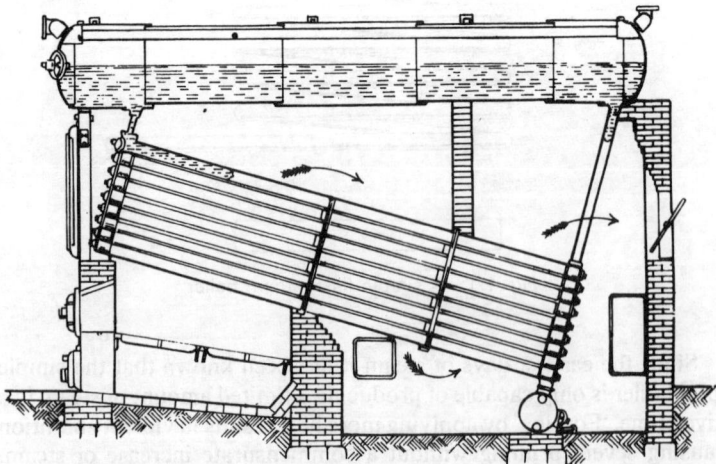

Fig. 1.2 An early Babcock land installation, of the longitudinal drum type

shell constructed of $1\frac{1}{2}$-inch-thick steel plates, and working at 200 *psi*.

Such boilers were heavy and bulky and the water-tube pattern found favour for many purposes, especially where light weight was a consideration, such as in high speed marine use. Both water and fire-tube boilers rely on convection or thermo-syphon action, operating by a difference of temperature—hot water rises and cooler water takes its place. The resulting circulation may be haphazard and poor in some designs. For example, poor circulation resulted in the dead water below the flue in some 'Cornish' boilers. For satisfactory action a good 'head' or height difference should exist between upper and lower sections—just as in any gravity hot water system used for domestic purposes.

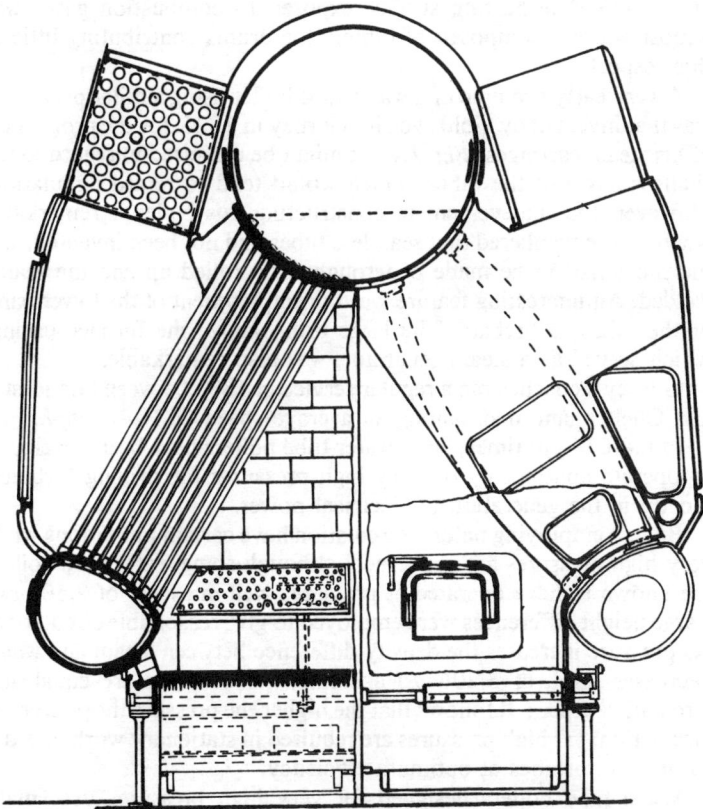

Fig. 1.3 A typical full-size Yarrow boiler. This particular example is fitted with air heaters at the side of the steam drum

This is reflected in several well established designs such as the 'Babcock and Wilcox' (*Fig. 1.2*) and the 'Yarrow' boilers (*Fig. 1.3*). Circulation in such boilers is extremely rapid, keeping the tubes free of scale, the latter being deposited in a 'mud drum' often situated at the lowest point. The water drums are much smaller than in the fire-tube type, thus allowing them to be thinner and light in weight, yet they can safely carry pressures up to 500 *psi*.

The 'Yarrow' pattern was often used for marine work as it represented a high output/weight ratio boiler, which was of suitable external dimensions for fast naval vessels.

Other designs of water-tube boilers employed several small drums of 30–36-inch diameter, some at the top and some at the bottom. Masses of tubes connected the drums, so arranged as to give good circulation. The heating surface exposed to combustion gases was almost entirely composed of tubing, the drums contributing little in this respect.

A very early example of a water-tube boiler using similar principles was that invented by Goldsworthy Gurney in 1825, for the propulsion of his steam carriages (*Fig. 1.4*). It might be criticised in regard to the shallow angle of the tubes, which would tend to retard circulation. However, the achievement in constructing this boiler is remarkable when it is remembered that seamless tubes had not been invented, and the tubes had to be made of wrought iron rolled up and the seams welded. An interesting feature was the employment of the lower bank of the tubes as firebars. The large drum above the furnace casing, which served as a steam separator, was also remarkable.

Gurney's coaches ran a regular service in 1831 between Gloucester and Cheltenham, maintaining an average speed of 10–12 *mph*.

In more recent times, large water-tube boilers have been developed to operate continously on very high pressures for driving turbines, mostly in the generation of electrical power.

Boilers employing natural circulation have certain limitations when very high pressures are concerned, although some water tube boilers are known to have operated at pressures well in excess of 2,500 *psi*. Large height differences were employed to give reasonable circulation. As pressure increases the density difference between steam and water decreases until, at 3,206 *psi* absolute, the densities are equal and circulation ceases. It follows that the higher the pressure the poorer the circulation, yet high pressures are required in stationary work in order to operate turbines at optimum efficiency.

Water-tube boilers designed for very high pressures use small-diameter steam drums, the mass of tubes providing the effective heating surface. Another problem is the separation of water and steam within the drum. Baffles may be used inside to assist separation.

I = Steam space
L = Water level
H = Upper drum
B = Downcomer

D = Lower drum
C = Lower tubes

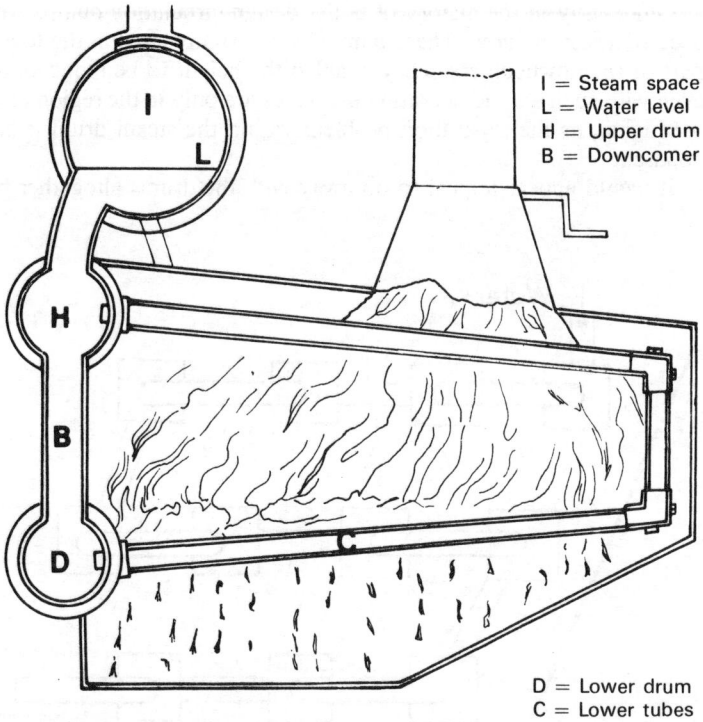

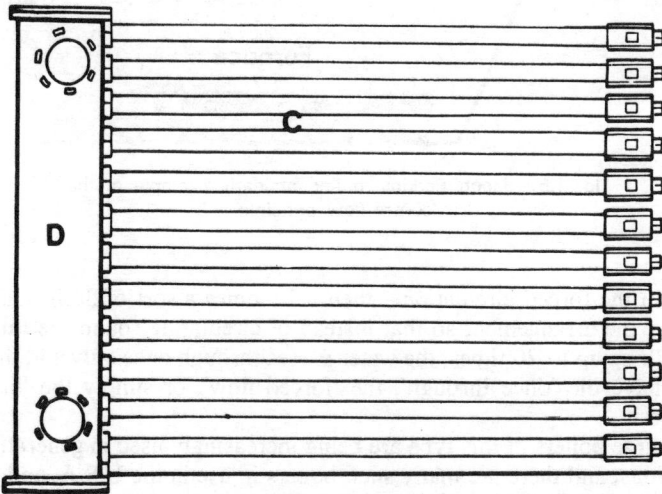

Fig. 1.4 The original boiler of the Goldsworthy Gurney steam carriage (1825)

Quite early in the history of boiler design circulation pumps were used to assist the flow. These pumps were usually sited in the lowest part of the downcomers. They enabled the height to be reduced, but even with such aid the operating pressures are only in the region of 50 *psi* higher, as the separation problem within the steam drum is still apparent.

It would appear logical to do away with the drums altogether but

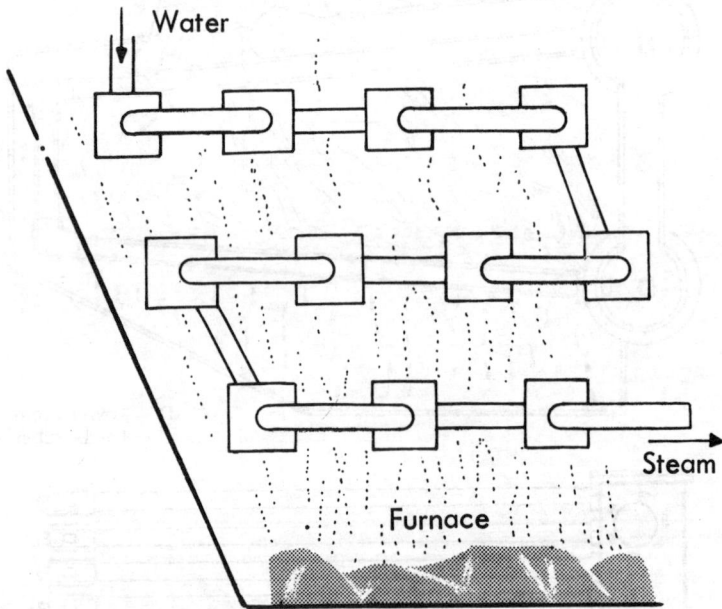

Fig. 1.5   Jacob Perkins' boiler, an early example of the 'contra-flow' principle

retain the forced circulation—then it becomes a sort of flash boiler. Alter the arrangement so that instead of circulating round the tubes anything up to 20 times, the water passes through once only and then we have the 'once through', the 'forced flow', or simply the 'flash boiler'.

Large boilers of this type are being increasingly used in generating stations, and there are many such boilers in use in the U.S.A. and on the Continent. It is understood that there are only two in use in the U.K. at present.

The two main types are the 'Sulzer' boiler, licensed by Sulzer Bros., and the 'Benson' boiler licensed by Siemans.

Boilers of more moderate sizes suitable for industrial use are made by Mitchell's of Peterborough. The availability of such boilers means that this is one field in which flash steam is no longer 'experimental'.

Although economical in material the construction of large 'once through' boilers is expensive, because of the complexity of the control system.

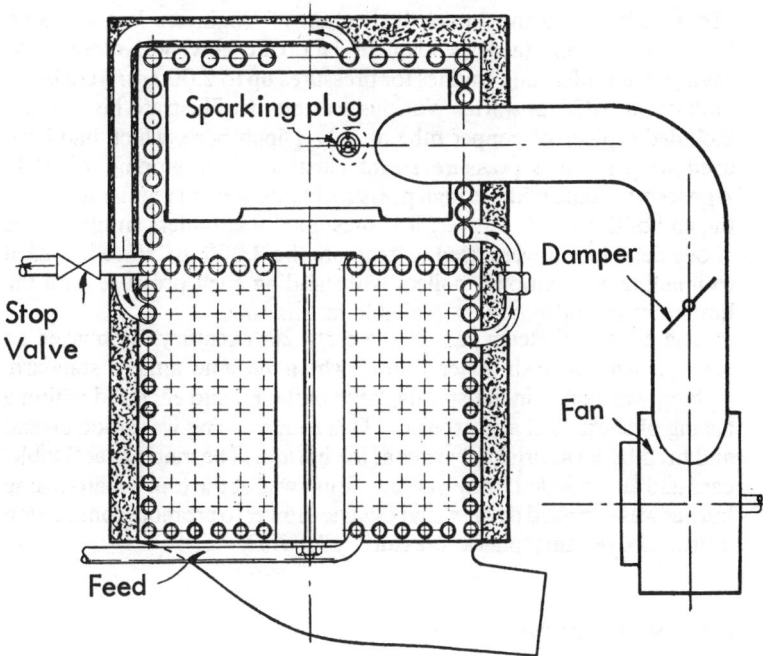

Fig. 1.6 The 'Doble' boiler

Temperature is regulated by control of heat input, pressure by control of feed water.

The time taken between water entering the boiler and emerging as superheated steam is a few minutes only. Therefore the control system must be sensitive and reliable.

Feed water used for these large flash boilers has to be very pure, as impurities can be deposited within the tubing or carried through. Since one boiler may develop tens of thousands of H.P. the water circulation is commensurately large, and impurities cannot be tolerated. Special methods have to be used to demineralise both condensate and additional feed water.

A very far cry from modern 'once through' boilers was the Jacob Perkins' boiler of the early 19th century. It consisted of a number of square section tubes made of cast iron connected in series and arranged in three layers over a furnance. Water was fed in the top end and steam taken from the end closest to the fire—here is an early example of the 'contra-flow' principle, of which more later (*Fig. 1.5*).

In an address to the Junior Institution of Engineers late in 1927, Loftus P. Perkins (a descendant of Jacob Perkins) was reported as saying that boilers and engines for pressures up to 2,000 *psi* were being built from 1822 onwards. Various exhibits to illustrate his address included a piece of copper tube of only $\frac{1}{8}$-inch bore, which had been used to pass high pressure steam capable of developing 40 H.P. Apparently some of these high pressure plants were in regular use up to the end of the 1914–18 war. It is presumed that boilers similar to the above description were used to generate the 2,000 *psi*, since it is hard to imagine conventional boilers withstanding such pressure with the limited range of materials available at this time.

The advent of steam cars in the early 20th century established an arrangement of flash-boiler tubing which became almost standard. Tubing was coiled in flat spirals, layer on layer, and enclosed within a casing of sheet steel and asbestos. This round casing was stood on end and fired by a vaporising burner at the bottom. The much later 'Doble' cars had an amended form whereby firing was at the top. An atomising burner was supplied by air via an electric blower to promote combustion a little above atmospheric pressure (*Fig. 1.6*).

Advantages of flash boilers

As mentioned earlier, high pressures and high output may introduce problems in certain types of steam boilers. If 'forced' by excessive heat, circulation may become intermittent due to large quantities of steam generated in the tubes; steam locking or reversal of flow can also occur. This cannot happen in the monotube flash-boiler, as the water is pumped under pressure greater than that of the steam. It is therefore possible to generate very large amounts of steam for a given heating surface without the above problems.

Another advantage of great importance is that of safety. When a boiler is under pressure the boiling point of the water is increased. Steam tables are available to show these temperatures, for example, at 100 *psi* gauge, the temperature of the water is about 170 degrees C.

This means that should serious failure occur in a conventional boiler all of the water is instantly turned into steam because of the sudden lowering of pressure. Steam occupies about 1,600 times the volume of water at atmospheric pressure, hence the boiler room is completely filled with scalding steam.

Flash boilers have little water capacity, so that should failure occur no vast amounts of steam are produced beyond the initial quantity. One of our colleagues was once leaning over a small flash boiler when a tube exploded—no injury occurred apart from his being startled by the loud bang.

The Board of Trade are very particular about the regular testing of all pressure vessels. Thus even some conventional model boilers might in theory be subject to test.

Some types of water tube or semi-flash-boilers are fairly quick at raising steam from 'all cold' but few can approach the flash boiler in this ability. Some examples of steam cars having electric ignition systems were claimed to start up in 30 seconds.

Control of the smaller flash boiler is also obtained by heat input and feed water. Since there is a small mass of material or water there is a correspondingly little residual heat capacity containing energy to oppose the change. Mention has already been made about the size and weight of conventional boilers. The flash boiler can be made very light. It can have a variety of layouts depending on the requirements. Cost *should* also be lower, because less time and material is required in construction (but very large 'once through' boilers are higher in cost at the present time).

In view of this formidable list of advantages, it is a pity that more use has not been made of its potential. There are, of course, disadvantages in some applications. The lack of water capacity might be serious in some uses—if the feed pump stops for any reason so does everything else! Uncontrolled flash boilers tend to superheat greatly when working at high capacity, and this again may be undesirable for some engines.

Applications

It would seem that flash boilers could displace a great number of conventional types. The truth is that there has never been a very large number in use at any time. Steam engineers appear to have avoided them like the plague! One disadvantage, not so far discussed, concerns the life of the tubing. Flash boilers appear to work best when pretty hot, and ordinary mild steel has a limited life under these conditions. The greater production of stainless steel tube in more recent times has led to the continuously welded type becoming fairly cheap, and it is to be hoped that the solid-drawn type may eventually follow suit.

Flash boilers have been used for steam cars, lorries and buses, cycles, small boats, aircraft and rail-cars (but not apparently for conventional full-size locomotives). Some of these were produced in small numbers, others were strictly 'one-off' affairs. As far as models' are concerned, boats have always been the most numerous, with a few locomotives and aircraft also represented.

In view of this relatively meagre use of the principle since its inception, flash steam is still in an experimental stage in most of its applications.

Pressures and efficiency

There has been considerable argument in the past concerning the pressures attained in model flash boilers, pressures being estimated to be anything from 25 *psi* to several thousand. These estimates although varying so drastically are all correct! Jim Bamford has recorded over 4,000 *psi* in a model flash plant, and at the other extreme, another plant showed a gauge pressure of only 25 *psi*. Both pressures were recorded under normal operational conditions. The fact is that the working pressure depends on the heat supplied, the amount of feed water and the steam consumption of the engine. The engine acts as a metering pump controlling the steam flow, and thus dictates the operating pressure for a given amount of heat and feed water. Abner Doble's famous steam cars worked at pressures of 750 *psi*, and developed up to 125 B.H.P. at very high overall efficiency—steam consumption was about 9·6 lbs. per B.H.P./hour using compound engines.

On a much smaller scale Jim Bamford has managed to record 20 lbs. per B.H.P./hour. This is a remarkable figure for a simple non-condensing engine only 1⅛-inch bore × ⅞-inch stroke.

It is believed that generating stations using 'once through' boilers achieve a ratio in the region of 6 lbs. per B.H.P./hour. But plenty of space is available for all possible economising devices, such as feed-water heaters. In addition, the prime movers are highly efficient turbines, which take steam at high pressure and expand it progressively in stages until discharged to a condenser.

When flash plants are operated on high pressures, efficiency also appears to be high. Uncontrolled plants, as used in models, also tend to superheat execssively, which may cause damage to the engine. De-superheating by water spray as used in some full-size boilers to deal with excessive superheat appears too difficult. Nevertheless, some form of heat exchange from steam line to feed water might be possible.

Certain early steam cars employing flash boilers without control of superheat were on occasion said to have cylinders glowing a dull red. They gave prospective passengers such a warm welcome that some of them refused to ride!

Superheated steam has long been renowned for its excoriating properties. An interesting theory was put to us by Prof. D. H. Chaddock to account for this. He suggests that the effect may be due to a chemical reaction producing an abrasive oxide. This is the action of steam on hot iron filings, which liberates hydrogen and black magnetic oxide of iron. The reaction commences at 200 degrees C and reaches commercial gas-producing proportions at red heat.

In his work with model flash steam turbines he found the small nozzles frequently became blocked by a mysterious brownish deposit. This happened in spite of blowing steam through the coil prior to each run. It was thought that the above reactions may have been the cause of the trouble, and if this is so it would account for the excessive wear sometimes reported on piston valves, etc. Steam separators, as used on some full-size plants, would probably help to overcome this if boiler tubing of mild steel is in use.

Because of the very high pressures that can be generated by flash boilers, it may not be out of place to mention the possibility of tube failure. From time to time formulas are quoted without reference to first principles, and a brief explanation may assist those readers who would like to know a little more about the subject, and also the reason why stress increases with diameter for a given pressure and wall thickness.

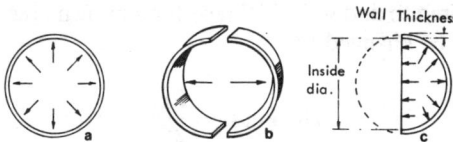

Fig. 1.7   Tube failure

Fig. 1.7 (*a, b,* and *c*) show cross-sections of a round tube. Steam pressure acts evenly on all parts as indicated by the arrows. Considering failure *along* the tube parallel to its axis, both walls are presumed to fail (in fact, one side would probably fail first, but this does not invalidate the principle). The diagram *b* indicates the two halves pushed apart. The total force exerted is equal to the area across the inside diameter (projected area) multiplied by the pressure (diagram

c). In this instance, area equals inside diameter times length. To give an example, using unit length to make things easier, on a tube $\frac{1}{2}$-inch inside diameter and pressure 1,000 psi, $\frac{1}{2} \times 1 \times 1,000 = 500$ lbs/F. This would be the force trying to burst the tube apart.

In mechanics, 'stress' is calculated by dividing the force (or load) by the total area resisting the force:

$$stress = \frac{force}{area}$$

The area resisting the force, in this example, is the area of the upper and lower sections of the tube wall = twice (thickness × length). Substituting this for 'area' and also (Pressure × diameter × length) for 'force':

$$stress = \frac{P \times D \times L}{2 \times T \times L}$$

When P = Pressure
L = length
D = diameter
T = wall thickness

Cancelling L, $$stress = \frac{PD}{2T}$$

Thus if 'D' is increased, the stress increases in direct proportion. The possibility of the tube failing around its circumference is of less concern, since a similar calculation proves that the stress is then equal to PD/4T—only half of the amount of the previous calculation.

If it is required to find wall thickness for a known pressure, the first equation may be changed to:

$$Thickness = \frac{PD}{2S}$$

when 'S' = permissible tensile strength of the material. Tabulated values for 'ultimate tensile strength' are not used here, since they are based on tensile tests to destruction when the material has been stretched beyond its 'elastic limit'. Working stress has to be much lower and because of this a 'Factor of Safety' is used, the value of which depends on the type of loading to which the material is subject. For example, a component subject to shock loads may need a Factor of Safety of 10, i.e., working stress will be only one tenth of the ultimate stress.

For boiler tubes the factor chosen may not be as high as 10, but is complicated by the fact that high temperatures lower the strength of the material and this must be taken into account. One reference quotes the working stress for copper steam pipes as 3,500 *psi*, and mild steel as 20,000 *psi*, but this for ordinary boilers where the tubes operate at a moderate temperature. Small diameter tubes as used in model flash boilers have an advantage—the small bore enables very high pressures to be contained without failure.

Simple application of the formula on a mild steel tube of $\frac{1}{4}$-inch bore and ·036-inch wall thickness (20 S.W.G.) will indicate that the bursting pressure is not reached in theory even at 14,000 *psi* (allowing 60,000 *psi* max strength). Nevertheless, tubes *do* fail at much lower pressures, for the simple reason that in a flash boiler temperatures exceeding 900 degrees C (bright red) may be commonly attained on some sections of the boiler coil. Some 'wasting' of the tube wall also may have previously occurred.

CHAPTER TWO

APPLICATIONS TO TRANSPORT

The early days of mechnical road transport were notable for the rivalry between steam-driven vehicles and those propelled by internal combustion engines. Some of the steam-driven cars were not 'flash steamers' since they employed boilers having some water capacity, which avoided some of the problems associated with flash boilers.

The famous Stanley steamer employed a fast-steaming fire-tube boiler with a two-cylinder double-acting engine geared to the rear axle. The pressure used varied, but in practice could operate at over 500 *psi*. The boiler has been quoted as not being entirely satisfactory, replacement of tubes being the main trouble. Firing was by a circular vaporising burner sited beneath the boiler.

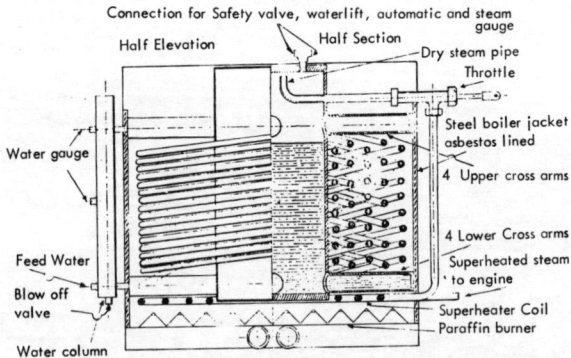

Fig. 2.1 Part of the sectional elevation of the Bolsolver boiler

This car was made in fairly large numbers, and it achieved a speed record of 127 *mph* in 1906—a performance which is still quoted by steam enthusiasts as a pointer to steam car potential.

Later, the Stanley concern was taken over by the American Steam Car Company, and in some cars the boiler was replaced by the 'Derr' water-tube boiler. The latter was composed entirely of $1\frac{1}{2}$-inch steel tubes about 20-inches long, welded together, no less than 100 tubes being used. The burner was of the atomising type, electrically driven and ignited by sparking plug.

Another replacement for the 'Stanley' and other steam cars was the semi-flash pattern designed by the Bolsolver Brothers and described in *Model Engineer* in August, 1927 (*Fig. 2.1*). Apart from the central drum the arrangement was similar to flash boilers used in many steam cars.

Plate 1   A fairly recent picture of Mr. Alec Hodsdon in the driving seat of his restored 'White' steam car

Some of the early flash-boilered steam cars included the 'White', the 'Serpollet' and the 'Turner-Miesse'. A vertical twin-cylinder compound engine was used in the 'White', the flash boiler being arranged in flat coils. A fairly elaborate diffused-flame vaporising burner provided the heat for this. Although the water feed pump was driven directly, a feed regulator controlled the water supply. One 'White' owner criticised the direct-driven pump claiming that the car tended to slow up on hills towards the end of ascent.

Plate 2   M. Léon Serpollet with one of his 1904 steam cars

Desirable features

Let us now state some desirable features for steam driven road vehicles.

1 Safety
2 Quick starting from all cold
3 Flexibility in traffic and hill climbing
4 Good acceleration, performance and range
5 No manual controls of the steam plant
6 Low running and maintenance costs

1 Safety—there appears no reason why a flash-boilered vehicle should not be as safe as any *i.c.* engined car. The fuel used would probably be less volatile, and since combustion in a modern flash generator is completely enclosed fire hazard need be no greater. Boilers having water capacity might be less favourably regarded although quite safe if properly constructed. Regular Board of Trade inspection would be required, which is obviously a disadvantage.

2   Starting up from all-cold can be achieved in under one minute with 'flash'. A total of $3\frac{1}{2}$ minutes has been quoted for the Bolsolver semi-flash type.

3   If the engine is the traditional D.A. slide or piston-valve type, geared directly to the rear axle, the engine will not be running when the car is stationary. Acceleration from rest is no problem if the engine has some kind of link motion to vary valve timing and sufficient supply of steam. With the flash boiler, the latter requirement is more difficult to meet, since there is virtually no steam reserve, and it is essential that the control gear operates instantly to enable the boiler to supply steam as the throttle is opened.

Hill climbing, too, presents problems; power developed must be high, yet the engine, if direct drive is used, will be running more slowly, probably with different valve setting, thus requiring more steam. This means that the water-feed pump ought to be separated from the *rpm* of the engine, but related in some manner to the power requirements.

4   The torque of steam engines at low speed is well known to be superior to *ic* engines, hence no gear box need be employed. Thus, if the steam is there the car will accelerate in a smooth progressive motion.

Performance of some of the old steam cars was stated to be only fair when compared with petrol-driven cars. However, since the power of a steam plant is governed by the boiler and heat supply the performance of some cars might have been very much improved with really good boilers and firing methods.

Range is tied up with steam economy—the less steam used the less water to condense. To quote extreme examples, a poor non-condensing steamer did only one mile per gallon of water. Yet the very efficient 'Doble' plant fitted in a German commercial vehicle required a $2\frac{1}{2}$- gallon reserve tank only, because of the large condenser permitted by the available space. A car operating at high pressure and fairly high temperature, with compound engine and condenser, might be able to cover over 200 miles on quite a small reserve water tank.

5   Some early steamers required a lot of hand pumping, and control of valve gear, etc. But it is quite possible to make everything automatic, as Abner Doble proved. Steam enthusiasts might not object to some hand controls, but for the general user control must be as simple as that for an ordinary car.

6   Most steam cars of the past were heavy on fuel when compared with some petrol-engined cars. The cheaper cost of lower grade fuels could balance this expense.

If well designed, incorporating modern materials, the steam car need be no more difficult for maintenance than any other vehicle, providing a sensible layout were provided with reasonable access to working parts.

The 'Doble' steam car

Practically all the points quoted above were met by the famous 'Doble' car last made about 1931. It must be emphasised that this car was the 'Rolls Royce' of steam cars and was far from cheap, £1,500 being quoted for the chassis alone. This is quite understandable for a vehicle of this class built in small numbers. *Plate 3* shows a type 'F' Doble owned by Thos. Hindle, Esq., of Blackburn.

Plate 3   The Doble steam car F-34, owned by Thos. Hindle
of Blackburn

The car was of very strong construction and ready for the road weighed 4,300 lbs. Indeed, the engine and rear axle unit together with the massive brake drums were intended to be interchangeable for use with commercial vehicles (*Fig. 2.2*) For the type 'F' Doble car the engine was a four-cylinder double-acting compound. The H.P. cylinder had a $2\frac{5}{8}$-inch bore, the L.P. $4\frac{1}{2}$-inch bore—both with 5-inch stroke. The crankcase was bolted to the rear axle casing, the engine driving the differential via spur gears.

Two piston valves only were required, one valve controlling one high and one low pressure cylinder, including admission transfer to L.P. cylinder and also exhaust from L.P. cylinder. Valve gear was

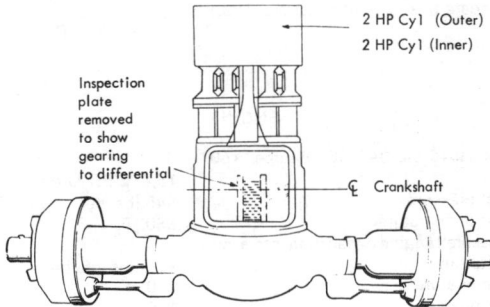

Fig. 2.2 Sketch of the 'Doble' engine and axle

Stephenson link motion, providing reverse and three cut-off points; 80 per cent for starting from rest and steep hills, 60 per cent for normal traffic conditions and 40 per cent for maximum speed and economy. These cut-off points were controlled by foot pedal which also gave reverse; accidental engagement of reverse was prevented by a safety latch.

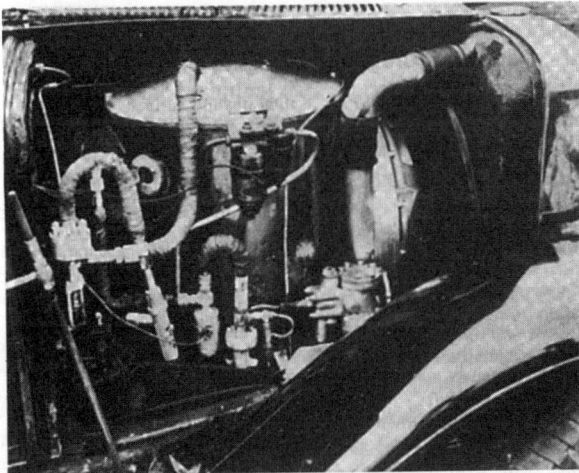

Plate 4   The Doble F-34 boiler

Table 1. Data on the Doble steam plant

DYNAMOMETER TEST ON DOBLE AUTOMOBILE STEAM POWER PLANT
(TEST No. 212.)

This test was conducted on August 9, 1929, with the same engine, after 72,300 road miles, engine No. 27, that was used in test of Test No. 198, after 46,200 road miles. (Compare this test with Test No. 198.)

ENGINE

Fuel—Shell stove-oil, 34°–36° Baumé, 150° flash	19,610 B.T.U./lb.
Steam chest pressure . . .	845 lbs. per sq. in.
Steam chest temperature	880° F.
Exhaust pressure (engine exhausting into a low-pressure turbine)	11·5 lbs. gauge.
Exhaust temperature	320° F.
Shaft horse-power	125·0.
Shaft speed	
Cut-off (in all cylinders)	40 per cent.
Steam per hour	1,201 lbs.
Water rate	9·6 lbs./shaft H.P. hr.
Fuel rate	0·916 lbs./shaft H.P. hr.
Heat liberated in furnace	17,950 B.T.U./per shaft/H.P. hr.

DYNAMOMETER TEST ON DOBLE AUTOMATIC STEAM POWER PLANT
(TEST No. 198.)

The test was conducted on June 16, 1928, with the standard power plant in the Research Laboratory of Doble Steam Motors, Emeryville, California. The following data were recorded:

BOILER

Standard type, containing 88·0 sq. ft. total heating surface, with type 5 furnace of 3 cub. ft. volume; test was run without feed-water or combustion air-heater.

Steam pressure at throttle	850 lbs. per sq. in.
Steam temperature at throttle . . .	850° F.
Actual evaporation	1,173 lbs. per hour.
Evaporation from and at 212° F . . .	1,678 lbs. per hour.
Fuel consumption	104·6 lbs. gasoline per hour (20·570 B.T.U. per lb.).
Forced draft motor input	980 watts.
Evaporation from and at 212°F. per lb. of fuel	16·03 lbs.
Overall boiler and furnace efficiency . .	77·75 per cent.
Feed water temperature	75° F.
Flue gas temperature with shrouded thermo-meter (error unknown)	518° F.
Air temperature entering burners . .	80° F.
Air pressure entering burners . . .	5·8 inches of water column.
Heat liberation in furnace	716,000 B.T.U. per cub. ft. per hr.
Heat input at burners	2,150,000 B.T.U. per hr. = 17,976 B.T.U. per H.P. hr.

ENGINE.

Four-cylinder compound; two high-pressure cylinders, 2¼" bore; two low-pressure cylinders, 4½" bore, 5" stroke; 40 per cent. cut-off in all cylinders. Engine No. 27, after 46,200 road miles. New piston-rings installed at 41,000 miles.

Steam chest pressure	710 lbs. per sq. in.
Steam chest temperature	826° F.
Exhaust pressure	1¼ lbs. gauge; 15·95 lbs. per sq. in. absolute.
Shaft horse-power	119·6.
Shaft speed	937 revolutions per minute.
Water rate	9·8 lbs. per H.P. hr.
Fuel rate	0·874 lbs. per H.P. hr.
Heat input	139·70 B.T.U. per H.P. hr.

The success of the Doble car was largely due to the flash-steam generator and to the patented feed water control system, which maintained constant steam temperature quite automatically regardless of load or changes of load. Diameter over the casing was 22 inches. Height was 40¾ inches including furnace and flue. Heating surface was quoted as 74 square feet. Weight was 484 lbs. with burner. Evaporation 1,330 lbs. per hour at 800 *psi* and 800 degrees F (427C). *Plate 4* shows the boiler of Mr. Hindle's car.

With regard to firing: a 12-volt electric blower was used to blow air through the venturi, drawing fuel from a float chamber. The mixture was ignited by a sparking plug and coil ignition system.

Those who like some technical data on performance should refer to *Table 1*, published with the above information and other data in *'Model Engineer'*, February 6, 1930. Evidently it refers to a slightly different plant as engine bore is stated as being 2¼ inches, and heating surface 88 square feet.

Reverting to some of the earlier vehicles, the Serpollet car was also a flash steamer, but the engine was similar to *ic* engine practice rather than to the more conventional steam layout. The illustration (*Fig. 2.3*) shows the arrangement of the 90-degree, vee-four engine; cylinders were single acting, with poppet valves for control of inlet and exhaust. The four valves in each block were located in a manifold and operated by a sliding camshaft. It is understood that the engine was directly coupled to the driving wheels. The flash boiler employed was of heavy gauge steel tubing fired by a vaporising type of burner.

A Serpollet steamer of special construction was stated to be the first road vehicle to attain 60 *mph*. This performance was officially recorded on the *Promenade des Anglais* at Nice in 1899.

Mons. Serpollet also designed and constructed steam tramcars which ran in Paris from about 1893 onwards. Further development of vehicles was terminated by the early passing of this brilliant engineer.

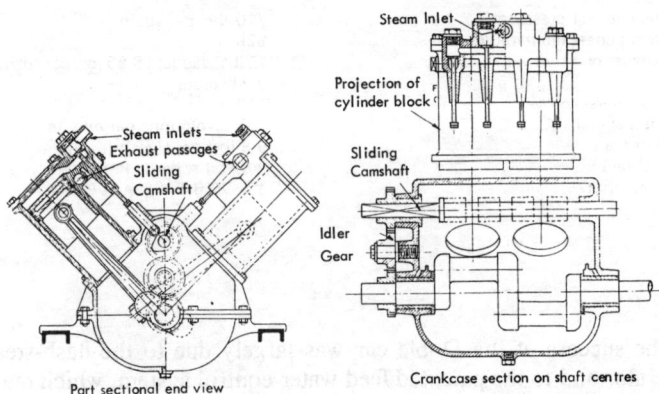

Fig. 2.3   Serpollet V-4 steam car engine

The late Sir Robert Bland-Bird said that he possessed a two-seater Serpollet car in 1903, which on one occasion was driven from London to Glasgow to take part in a hill-climbing competition. This journey was accomplished without trouble—a remarkable performance in those days. A boiler failure, however, prevented a win when approaching the summit of the hill climb.

In the early years of this century steam omnibuses were a feature of the London scene, and The Metropolitan Steam Omnibus Company operated some Darracq-Serpollet buses which had flash boilers. Another fleet of steam buses run by the National Steam Car Company is better known, probably because it operated on a larger scale.

In view of recent reports and speculation in the press about a revival of the steam car, it is not out of place to refer to this possibility. At the present time a great deal of public anxiety has been expressed on the subject of pollution of various kinds. As far as air pollution is concerned the conventional petrol-engined car has come under fire. This is mainly in the U.S.A. and Scandinavia, where large numbers of cars tend to be concentrated in urban areas.

Apart from extreme proposals to ban petrol engines entirely; in the U.S.A. legislation is already due to come into force in 1975 limiting exhaust emission of certain gases. This has caused motor manufacturers to examine the possibilities of the steam car, and a great deal of money and time has been spent by several famous motor firms. The theoretical reasons for this interest lie in the fact that external combustion, if properly carried out, produces a mere fraction of the pollution of the petrol-engined car not fitted with special exhaust devices.

At the time of writing no steam cars have yet left the production line, but it is known that several experimental cars have been built, for example, a press report in November, 1971, discussed a steam 'minicar' being developed at British Leyland by Sir Alec Issogonis.

A most interesting paper by R. M. Palmer was read to the automotive section of the Institution of Mechanical Engineers in 1970. Entitled *An Exercise in Steam Car Design* it dealt extensively with steam car problems. The original work was on behalf of Ricardo's of Shoreham. The research division of General Motors instigated this research, requiring an independent study to run parallel with their own.

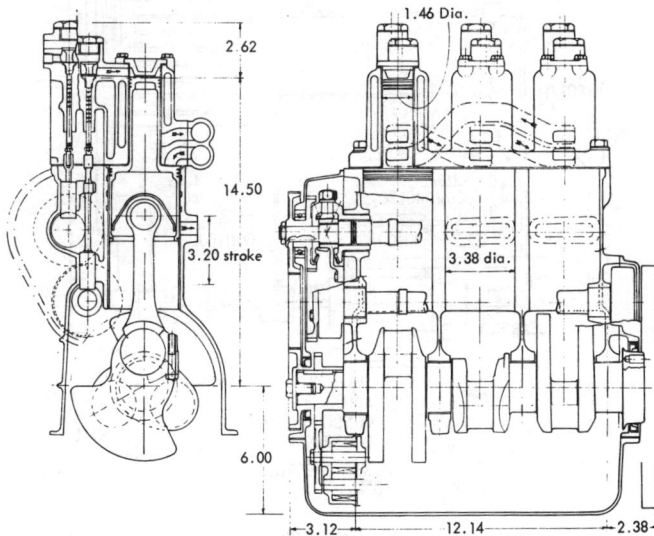

Fig. 2.4   The L-type, three-cylinder steam engine

Apart from the theoretical discussion of requirements, a complete steam-car plant was designed to meet certain specifications.

The conventional direct-geared engine was avoided for a number of reasons, and there appeared to be advantages in the ability to declutch the engine from the transmission. A three-cylinder, single-acting cross-flow tandem compound engine (*Fig. 2.4*), with uniflow exhaust from both high to low pressure cylinders and from low pressure to conden-ser, was designed to run at 2,000 *rpm*. Power aim was 60 B.H.P. as supplied by a flash boiler capable of evaporating 700 lbs./hr. at 1,500 *psi* and 900 degrees F.

In order to obtain variable cut-off with poppet valves, two valves in series were envisaged, one of which could be phased to open earlier (without steam admission, as the first valve would still be closed).

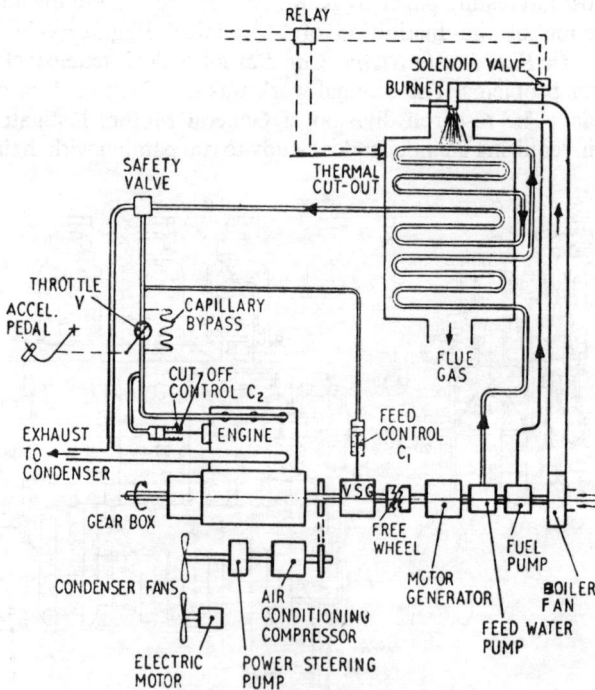

Fig. 2.5 Arrangement of controls and auxiliaries from R. M. Palmer's design (proceedings of the Institute of Mechanical Engineers, vol. 184).

Closing would also be earlier thus providing an earlier cut-off.

The control system for this plant is shown in *Fig. 2.5* and it will be seen that power for auxiliaries is taken from the engine—a feature made possible because the engine can revolve with the car stationary.

Of immense interest to steam enthusiasts are some of the conclusions of this research, summarised as follows:

1   The steam plant can not compete on technical or operational grounds with conventional ic engined cars, however, if exhaust emissions are an overriding consideration, a practical steam car does seem to be a possibility.

2   A fully condensing steam plant could be installed in the bonnet space about equal to that in a conventional car with top speed of 80 mph, and comparable acceleration to petrol engined cars. Fuel consumption better at speeds up to 30 mph, but worse in the 50–80 mph range. Cheaper fuel could be used.

3   Steam appears the best working fluid—others have insufficient stability at high temperatures. Fluorocarbons—which appear to be an alternative, give off toxic products if fluid escapes into the combustion zone.

4   Exhaust emissions of a steam plant should be favourable.

5   Using steam, high pressure is necessary for economy, consequences of a major leak are not serious if quantity of fluid in circulation is small.

6   An operational disadvantage is the freezing problem—no suitable anti-freeze additive has yet been found.

7   The steam plant is more bulky than the petrol engine equivalent yet gives lower performance. Control system is complex, and the cost of the plant likely to be greater than the conventional petrol engine plant.

We are indebted to the author, R. M. Palmer, and the Institution of Mechanical Engineers, for permission to publish the diagrams shown in *Figs. 2.4* and *2.5* and the above extract.

Railways

No conventional locomotives appear to be on record that have had flash boilers. By 'conventional' is meant locomotives capable of hauling heavy loads over long distances. However a well-known contributor on full-size locomotives, Chas. S. Lake, reported in *Model Engineer* in 1937 on a 'Besler' steam railcar put into service by the New York, New Haven and Hartford Railway Co.

The railcar was of the two coach pattern and streamlined in appearance; two existing coaches being modified to make up the set. To quote Chas. S. Lake "The principal interest lies in the propelling machinery, which consists of two two-cylinder compound engines, high pressure

$6\frac{1}{2}$-inch, low pressure 11-inch diameter, 9-inch stroke. These cylinders drive on to crank pins in the wheels, the mechanism being enclosed. The engines are of the double-acting pattern with piston valves, the crossheads being cylindrical in shape and made of cast steel. Designed steam pressure 1,500 *psi* and at 1,200 *psi* inlet pressure the power bogie has an average starting tractive force of 15,000 lbs. The motor truck is rated at 1,000 H.P., although capable of producing more than this with sufficient boiler capacity (*Fig. 2.6*).

Fig. 2.6   Besler steam motor train, New York, New Haven and Hartford Railway

"The boiler is of the continuous-flow non-water level type. It has no drums or headers, but is a continuous tube from the boiler inlet to the throttle. The water enters the top and passes down through a series of flat coils, where it is heated and then boils in the helical coils at the bottom surrounding the combustion space, afterwards passing to the superheater-coils just above the firebox and emerges as superheated steam. The boiler is 4 feet in diameter, and 6 feet 5 inches in height. The oil burner is of the pressure atomising type of special design and construction.

"The train performs a total daily mileage of 317·26 miles and operates in almost continuous service from 6 *am* to 10.20 *pm*."

It would seem that this railcar with its 1,000 H.P. rating is probably one of the most powerful flash plants on record as applied to transport. The description of the operation of the flash boiler could hardly be bettered, seemingly following the pattern of successful steam-car flash generators but on a much larger scale. No details of running costs are given, but in view of the fact that in 1937 Diesel engines were well established for shunting purposes: it could be inferred that at this time costs were not too disproportionate.

Lovers of the steam locomotive will no doubt speculate on developments that might have happened! One of the factors quoted in the late steam versus diesel controversy was the fact that steam locos required special steam coal, and had to keep fires banked up for long periods while off duty, otherwise considerable time was taken to raise steam from all cold. At least the flash boiler could have bettered this!

Other flash steam transport

The thought of riding a bicycle driven by a flash plant would not perhaps appeal to the timid yet there were at least two commercial steam bicycles, the Turner-Miesse (1902) and the Pearson Cox (1912/14). Also several amateur built jobs are on record. The attractions of such a machine were mainly for the enthusiast and it seems that not a large number were actually built. Similarly a number of amateurs have fitted flash plants in full-size boats with a fair amount of success. One such craft was built by Bert Squire, of the Malden S.M.E., some time ago. The engine was a 3-cylinder, single-acting uniflow job fitted with poppet valves for admission only. The bore was $1\frac{1}{4}$ inch diameter with a stroke of 2 inches and the exhaust was arranged to discharge into a keel condenser. *Plate 5* depicts the plant on test with the engine mounted on bearers. It may be noticed that the vacuum gauge sited on the lower right hand of the instrument panel is registering zero, although there is 200 *psi* on the steam gauge. This is because the plant on this occasion was not installed in the boat, and no alternative condenser was available.

Feed water and condenser pumps were both of the 'twin' pattern, and attached one on each side of the crankcase. Drive was by a cross-shaft geared down 5 : 1 from the crankshaft. An interesting feature is the raising of the cylinders a little above the crankcase by means of distance pieces. A system of baffles assisted in the clearance of condensate via the gaps when starting up so that no water entered the crankcase and also prevented overheating of the crankcase section.

The boiler consisted of a total length of 144 feet of chrome alloy tubing arranged in a series of 'grids' about 14 inches square sur-

Plate 5   The three-cylinder poppet valve engine of the
steam launch *Sphinx* by Bert Squire

rounded by a square boiler casing. All boiler joints were welded and tested to 3,000 *psi* by hydraulic pressure. Boiler firing was by means of a multiple burner of the vaporising pattern with a lever-operated variable jet. Fuel consumption was in the region of $\frac{3}{4}$ gallons per hour at normal cruising speed. *Plate 6.* shows the boat in action during an

Plate 6 The *Sphinx* achieves 10 *mph* on a test run

early test run at a speed of 10 *mph*. This hull was about 20 feet in length and 5 feet 6 inches beam. Displacement was $1\frac{1}{2}$ tons. A three-blade propeller of 11-inch diameter × 8-inch pitch was employed and the engine turned this very easily. With a steam pressure of 500 *psi* the boat was too fast for the Thames Conservancy speed limits, and it was found that a pressure of 350 *psi* was quite sufficient for all normal purposes.

The engine was really quite small and when visitors from the British Light Steam Power Society visited this craft one member was very surprised on being shown the engine—he thought it was the bilge pump!

Most modern tractors use diesel engines for motive power but during the Second World War some development was carried out on flash steam by R. H. and H. W. Bolsolver and Messrs. Rodgers and Geary Ltd. A joint patent was obtained for a system by which a conventional tractor could be converted to steam using a coke-fired flash boiler and modification of the i.c. engine to steam. A full account of this was published in *Steam Car Developments and Steam Aviation* which was later reprinted in *Model Engineer,* November 30, 1944.

It was claimed that 42 B.H.P. was developed as against 37 B.H.P. with reduced running costs. Not a great deal was published on the project after the date mentioned but a reader suggested afterwards that some trouble had occurred with erosion of the boiler tubing due to the fact that it was in direct contact with the hot coke.

CHAPTER THREE

HOME-BUILT STEAM CARS

At some time or other most steam enthusiasts have dreamed of building their own steam car, but in the end very few have actually been commenced let alone completed. This chapter is not intended to instruct the reader 'how to make' nor, indeed, to denigrate such a project. Instead, we shall try and discuss a few of the requirements—particularly as applied to model engineers or amateurs with limited equipment.

It would perhaps be as well to commence by discussing equipment and facilities. It is pointless even to think about such a project if, for example, there is no lathe or insufficient space available.

To be able to work on a car of average size requires a good deal of space and those who possess lock-up garages of the popular 14 feet × 9 feet size will be aware how difficult it is to work on a car inside—there is just not enough room to get round properly. Steam car construction would probably be a long term affair, so some kind of shelter is surely a necessity.

Some full-size boats built in back gardens have been protected by frames covered with thick polythene sheet and this seems the only alternative to a garage. Although such temporary constructions offer some weather protection, they are far from ideal. In very cold weather portable heating devices are usually insufficient to maintain a temperature suitable for work even for the toughest enthusiast.

Assuming that work space is available, we must now consider shop facilities. Many amateurs' workshops have a $3\frac{1}{2}$-inch lathe as their main machine tool—this represents about the minimum size capable of tackling the machining, and even then some work may require a larger lathe. A drilling machine is also high on the list of machine tools, and most shops will have these two essentials. If a shaper or small milling machine is available, a great deal of hand work can be eliminated, although some milling can be done on the lathe. Besides a good selection of workshop hand tools the usual garage pattern spanners and wrenches would be required. A great number of model engineers possess the equipment so that the problem of working area is the real criterion.

Now looms the question: how ambitious is the project to be? There are three options which may perhaps be listed: *1   To make a vehicle from scratch including plant, chassis, and body. 2   To convert an existing vehicle, making the plant but altering little else. 3   Combinations of 1 and 2 in varying degrees.*

To manufacture a vehicle completely is an ambitious project indeed, but not beyond the capabilities of some amateurs. The word 'completely' is perhaps incorrect for it is pretty obvious that many standard car components would be used in building up a modern vehicle. Chassis work would probably require welding equipment unless a return to some of the earlier methods of chassis construction were used. Amateur-built bodywork has often *looked* 'amateurish', but fibreglass has now made it possible to build up complicated shapes if correct methods are used. A design that was intended to be built from scratch is shown in *Fig. 3.1.*, and although many years have elapsed since its

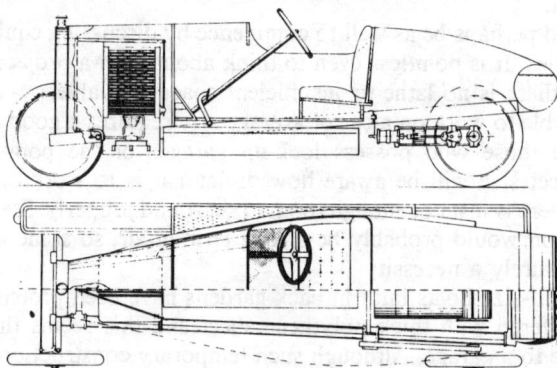

Fig.3.1   The light steam car design by W. S. Paris (1932), showing the part sectional elevation and plan

publication the proposed layout may be of interest. It was done in 1932 by W. S. Paris. It will be noticed that the chassis arrangement is based on two steel tubes.

The double-acting engine (single-cylinder, in this instance) is arranged horizontally to drive the rear axle via spur gears, with the boiler arranged as depicted. This, of course, is the classic layout of many early steam cars, and of the Doble (which was being manufactured around this time).

One of the advantages with this arrangement is that the whole of the bonnet space is available for boiler and condenser, with the engine nicely tucked out of the way. A disadvantage often quoted is the fact that if the engine is attached to the axle casing, the weight is unsprung. Designers might get around this by employing a 'swinging arm' type of suspension and a chain or modern vee belt drive to the axle.

Another factor is the size of the engine. Most engines arranged directly on the axle casing were fairly large ones and manufacturing problems arise if only small machine tools are available. This is not to say that precedent need be exactly followed, as larger gear ratios between engine and differential would allow a smaller but higher revving engine.

Chain drives were used on some early vehicles, but sometimes caused 'snatching' and a differential has to be worked in somehow. This is perhaps less easy than in designs where the engine is bolted directly to the axle casings. There are several possible basic layouts, it is not essential to have the boiler sited under the bonnet. Flash boilers are very accommodating, and can be placed in a variety of positions, updraught, downdraught, and what have you. If it was desired to use a standard car rear axle, complete with differential an engine position in line with the propeller shaft would be used; but it is by no means essential to have the engine under the bonnet especially if the engine design allows under floor mounting. The universal joint on a modern prop shaft allows a drive to be transmitted over a fair angle and a shorter shaft would not matter much.

Now to consider the opposite extreme—the conversion of an existing vehicle from ic to steam with as few alterations as possible. Query number one—what type of car! It is hardly likely that a new or nearly new car would be the subject of experiment. thus, a sound secondhand vehicle must be found. The 'Achilles heel' of most modern cars is the underbody rust problem, and in view of the work involved in a flash steam conversion, a car is required that is likely to last a reasonable time. What a pity that the separate chassis is out of fashion! Only a few mass produced cars in the past 20 years or so have been provided with a chassis and since the bodywork of such vehicles can be completely removed, advantages in steam conversions are obvious.

The most direct method of conversion is to design an engine to fit directly to the clutch bell-housing, using the original flywheel and clutch. This arrangement materially affects the design of the engine: it must be capable of fairly high *rpm*, as the rear axle ratio is usually around 4 or 5 : 1 for most cars with engines forward and using rear-wheel drive. We have not forgotten transverse engines and front wheel drives but consider such layouts less suitable, partly because of the small space available for boilers. An engine in the boot, driving rear wheels via a gear box might have distinct possibilities.

With an engine attached to the bell housing, double-acting jobs give way to single-acting cylinders. If the boiler is to be accommodated under the bonnet as well, it becomes necessary to keep the engine short—some conversions of this type have had a vee-twin or radial design using the minimum amount of space. Examination of contemporary cars shows that low ground clearance is against boilers situated below floor level. A boiler in the boot sounds attractive but apart from using up valuable luggage space some alterations would be required in order to obtain adequate ventilation for both incoming air and flue gases.

There are two schools of thought concerning the type of engine and whether it should drive directly or through a clutch and gearbox. The purists maintain that since a steam engine can develop a large torque at low *rpm* and can be self-starting, gearboxes and clutches are quite unnecessary. They maintain they would spoil some of the classical features of steam cars such as silent running, and the fact that the engine is stopped when the car is stationary—as, for example, in the instance of steam locomotives. The other school point to undoubted advantages in easier conversion and also other factors such as:

(*a*)   *The engine may be smaller; thus a $3\frac{1}{2}$-inch lathe can be used for most if not all of the machining.*

(*b*)   *Engine can have a fixed cut off without the necessity for special valve gears, sliding cams, etc.*

(*c*)   *Self-starting not necessary as the normal electric starter motor may be used to crank the engine—hence engines having only two single acting cylinders can be used.*

(*d*)   *Since the engine runs all the time the car is on the road, engine driven feed pumps and auxiliaries are a possibility.*

It is quite likely that such an engine would not be as quiet as the low-revving direct drive engine. Somebody once remarked that if you build a steam engine on petrol-engine lines it tends to run like a petrol engine and vice versa.

Our friend and colleague, the late Edgar T. Westbury, was president of the Model Power Boat Association and also president of the British Light Steam Power Society, and it is well worth quoting his views as expressed in the *Model Engineer* in October, 1966, concerning his experiments in designing an engine suitable for light steam cars.". . . There has been far too much 'heap big palaver' and theorising over the real or imagined virtues of steam cars: much of it is romantic wishful thinking and some of it unrealistic nonsense. When I met Abner Doble before the war, he observed that any steam car intended to beat the modern petrol-driven car would 'have to be darned good'—and I

believe this to be even more true at the present day. But that should not
deter enthusiasts from taking up the challenge.

"Many of the ideas about steam cars which have been expressed in
this and other journals ignore the basic problem of power-weight ratio.
Heavy boilers and still heavier engines, have been advocated by many
theorists. The use of any boiler having a large steam and water capacity
is almost completely ruled out for a light road vehicle for several
reasons. First of all, its weight for a given steam generating capacity,
second, because of Board of Trade regulations for pressure vessels and
the equally complicated matter of getting the plant and complete
vehicle adequately insured, and third, but by no means least, any boiler
which relies on circulation by convection presents great difficulties in
design for use at high pressure and variable rate of output.

"The only boiler which is suitable for these conditions and light
enough to fit to a car is the forced circulation type, the only simple
example of which is the flash boiler .... The burner is perhaps the
greatest problem of all, as it must have a very wide range of control,
preferably automatic, and be capable of instant response to demands ...
As regards the engine with which I am chiefly concerned at present,
the orthodox type of double-acting slide or piston valve engine, though

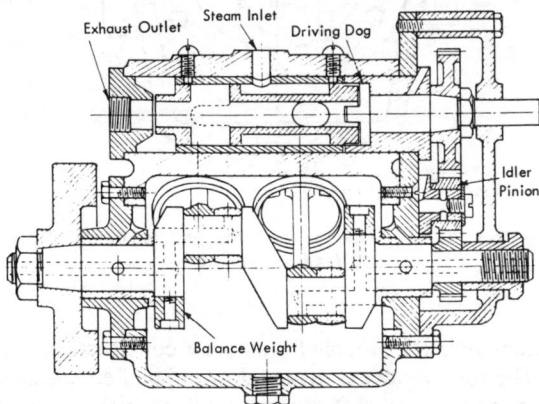

Fig. 3.2   The side elevation of the proposed vee-four rotary
valve engine by E. T. Westbury

very good if properly designed, would be very difficult to reduce to
light weight and small bulk. If run at high speed it is difficult to balance,
and the multiple parts of the valve gear tend to cause mechanical loss.
The alternative is a simple and compact single-acting engine *designed*

to run at high speed. Flexibility will be less, and auto-reversing may need to be sacrificed, so that the ideal of a gearless car may be difficult to realise . . . ."

Figs. 3.2 and *3.3* are two views of a 'Westbury' 4-cylinder, vee-type engine which was originally a preliminary design for application

Drain valve   Exhaust ports   Steam ports

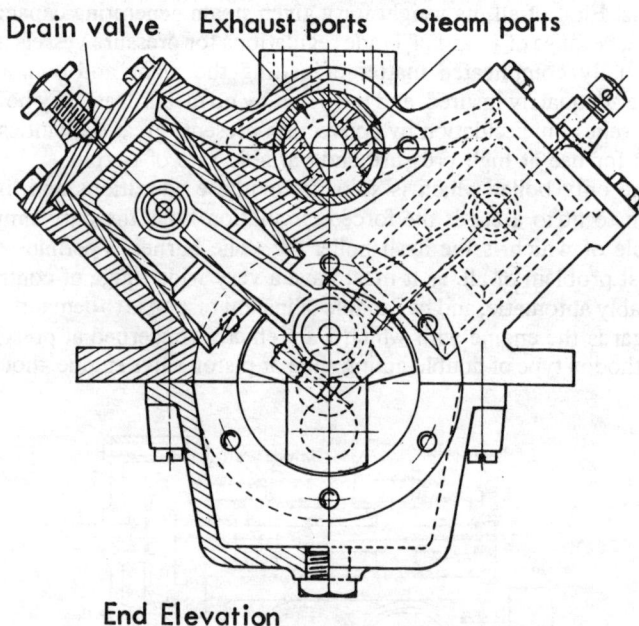

End Elevation

Fig. 3.3   Larger scale view of the end elevation of the 'Westbury' engine

to light steam vehicles suitable for amateur construction with limited facilities. The rotary valve is geared down to half engine speed and is driven by a tongue so that it 'floats' and is not subject to side thusts caused by the drive spindle. Cylinders are liners inserted in the monobloc casting. A two-bearing crankshaft was chosen in order to keep the engine fairly short, and the two-cylinder blocks are offset so that the big ends can operate side by side on the same crankpin.

It must be emphasised that no further work was done on the design as it was not approved by those concerned. However, the arrangment is compact and would appear to offer distinct possibilites.

Another problem facing the home constructor is the boiler and its control system. The pancake arrangement of tubes as per the 'White' and 'Doble' boilers would presumably be used as it is known to have worked well. Whether or not to choose top or bottom firing is another matter. A design by R. H. Bolsolver for a paraffin burner is shown in *Fig. 3.4*. This was intended to fire a small stationary boiler of the

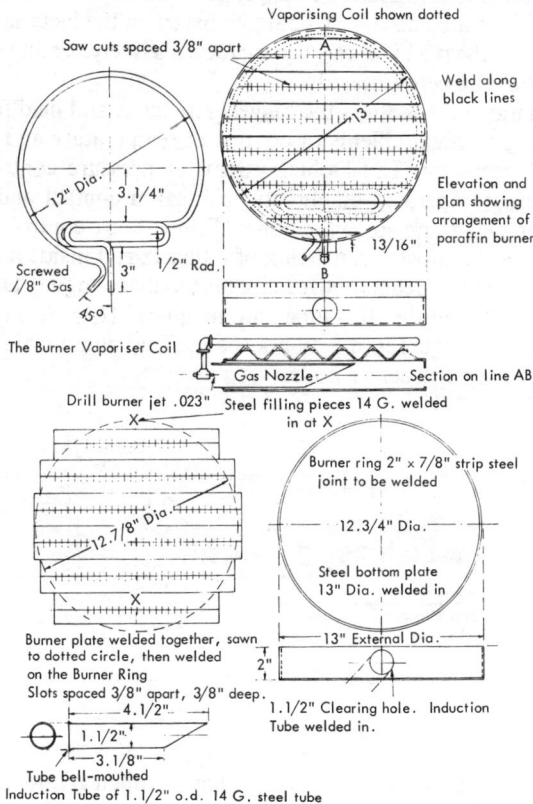

Fig. 3.4 Details of the burner design by R. H. Bolsolver

Bolsolver semi-flash pattern and all air for combustion was intended to be drawn through the induction tube. The flames produced were stated to be blue, tipped with orange, and the burner to operate very silently. Refer to *Fig. 2.1* for arrangement of the burner at the bottom of the casing. Two induction tubes are shown here for a larger boiler. The boiler illustrated is the semi-flash pattern not a flash boiler, and would be classed as a pressure vessel.

The cost of boiler tube is likely to be heavy if stainless steel is used—one Doble boiler used over 850 feet of tubing! It is not thought that as much as this would be necessary, as another Doble plant was capable of developing 125 S.H.P. from 88 square feet of heating surface but this, of course, with compound engines. This power ratio would represent 1·42 shaft horse power from 1 square foot. An amateur job is most unlikely to approach this ratio, but some idea of the quantity of tube necessary could be based on the fact that 8 feet of $\frac{1}{2}$-inch O.D. tube is a bit over 1 square foot, if a likely heating surface to power ratio can be estimated.

Automatic control offers the biggest challenge and field for experiment and involves problems of control of temperature and pressure, and quick response. The Doble system for pressure control used a steam pressure operated diaphragm to break a contact and stop the electric fan delivering air to the boiler. Temperature control was taken care of by a thermostat consisting of a quartz rod inside a steel tube heated by a loop of steam tube feeding the auxiliary engine. Differential expansion between the steel tube and the quartz (*Fig. 3.5*) operated a control lever bringing into action a second feed pump which injected water into the superheater coils.

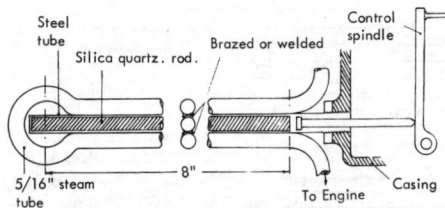

Fig. 3.5   The 'Doble' thermostat

When a condenser is used it has to have a much greater area than a conventional car radiator if all exhaust steam is to be condensed. There are possibilities of using commercial radiators which of course have a larger surface area, if space is available for fitting.

Writing in *Model Engineer* in December, 1956, George W. McArd described two steam cars of amateur construction. One was designed and built by Charles F. Keen in the U.S.A. and followed the classic layout of rear-mounted, double-acting engines, with the flash boiler mounted under the bonnet and an efficient condenser replacing the radiator of a petrol car. Combustion occurred at the top of the boiler and the flue gases arranged to be discharged at the bottom. No details as to size or length of tubing were quoted but the arrangement and layout were similar to the 'Doble' (*Fig. 3.6*).

Several different engines were tried, the third one being a modernised 20 H.P. Stanley of 2-cylinder, double-acting pattern 4-inch bore and 5-inch stroke.

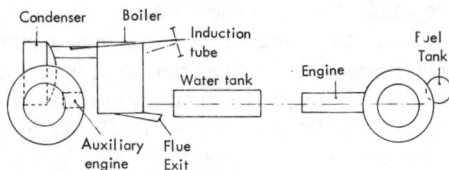

Fig. 3.6   The 'Doble' layout

Maximum pressure of 1,500 *psi* and temperature 800 degrees F (427 degrees C) could be used, both automatically controlled. To assist condensing, an exhaust turbine-driven fan was placed behind the condenser, and it was stated that in cold conditions condensing was so efficient that water loss was nil. Weight of vehicle in working order was 24 cwt. and top speed 75–80 *mph*.

The chassis and body of this car were built entirely by the owner and the only standard car parts used were details such as wheels, brakes, etc. A much later 'Keen' steamer is of the sports car type and very smart indeed.

Another steamer described in the same article was built by H. Claydon and although a 'Derr' watertube boiler was used and not a flash boiler, the engine appears of special interest as it could be almost completely built on a 'Myford' $3\frac{1}{2}$-inch lathe. The engine was quoted as a vee-twin 2-inch bore, stroke $2\frac{3}{8}$-inch, poppet valve inlet, uniflow exhaust, having fixed cut-off at 25 per cent stroke. A 3-speed and reverse gearbox was used, the engine being mounted on a plate to bolt to the bellhousing. This plate was the only item requiring the use of a larger lathe during manufacture.

Water and fuel pumps were mounted on the front of the engine and driven by reduction gearing—in fact, the engine is very reminiscent of model hydroplane examples but scaled up. As in the 'Keen' car the chassis and bodywork were home-built, the body panels being of aluminium sheet.

Although the performance was stated to be modest the car was fairly successful. It started up easily, since the engine could be run up to clear condensate and warm up.

Some years later this car was fitted with a home-built 3-cylinder, two-stroke petrol engine and Mr. Claydon wrote that the car had covered thousands of miles under steam before change.

Mr. McArd considered that a good type of engine for amateur construction was the four-cylinder radial pattern and this seems to be sound reasoning. The engine could be short and directly fixed to an existing bellhousing and gear box, or the gearbox perhaps dispensed with if variable timing was incorporated.

One great advantage would be the fact that a simple overhung crankshaft would be possible. Such a crankshaft is very much easier to produce on a lathe of limited size than a multi-throw crankshaft.

We are indebted to the British Light Steam Power Society for details of a modern home-constructed car by H.W. Kendrew of Windermere. This was reported very fully by Gerald Foster in a recent issue of the *B.L.S.P.S. Bulletin*. The following is an abridged description.

The car is a two seater having the front part of a Mini joined to a Renault rear axle. A chassis of steel tube joins the two halves. Engine is basically a 'Stanley', geared 1 to 1 to the axle. Exhaust steam from the engine passes into the tubular chassis member, inside of which is a copper tube carrying feed water. This arrangement acts as a feed-water heating system.

A flash boiler having 600 feet of tube provides 60 square feet of heating surface and is arranged in five coils around a central burner. The flame expands radially at first then upward through the coils and waste gases pass through exits at the top of the casing. The burner is fed with air/fuel mixture by a centrifugal fan. This burner sounds unique, and is Mr. Kendrew's own design. Instead of a position under the bonnet the boiler is placed behind the seats, and boiler control is by an expansion rod which operates a micro-switch. The latter in turn controls fuel supply—the auxiliaries are electrically operated.

The car is 114 inches long, with 47·5 inch-track and a 78 inch-wheel base. Cruising speed 50 *mph*, with occasional peaks of 60 *mph*.

An alternative possibility to amateur-built cars is the flash-steam motor bicycle—though not many of these appear to have been constructed. Such machines are not without their hazards—one intrepid builder related that in his experiments with a home-built sidecar outfit he ruined three pairs of trousers through catching fire. Asbestos suits are obviously at a premium in steam bicycle work!

There is obviously far less work involved in a bicycle project when compared with a car. The engine would be smaller and the main problem appears to be in the boiler, condenser and water tank departments, both in the matter of size and suitable installation within the frame. *Figs. 3.7* and *3.8* illustrate an interesting engine design for a steam bicycle by H. E. Rendall, published in the *Model Engineer* in

1960. Cylinder bore is $1\frac{1}{4}$ inches, and stroke $2\frac{1}{8}$ inches. Feed pump driven at half engine speed was $\frac{1}{2}$-inch bore by $\frac{5}{8}$-inch stroke. A simple slide valve was chosen to control both admission and exhaust. Mr. Rendall considered that excessive wear should not be a problem using modern lubricants.

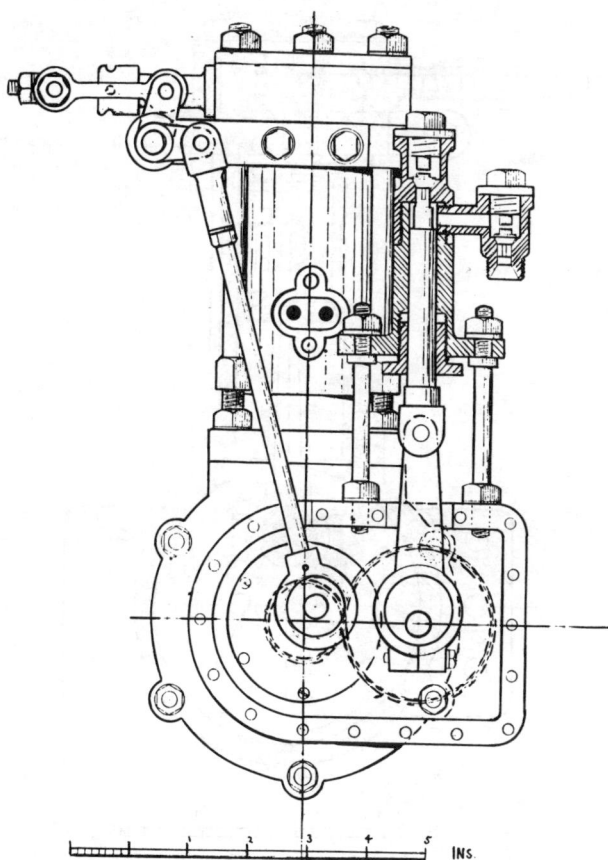

Fig. 3.7   Side view of H. E. Rendall's suggested engine for a steam bicycle showing the feed pump in section

From the foregoing it is clear that home-built steam cars are a practical possibility. Whether or not such a car could completely replace a conventional vehicle is perhaps another story; so much would depend on the degree of reliability attained and the performance required.

Mr. Alec Hodsdon, who has owned and rebuilt many steam cars and is regarded as having more practical experience of steam cars than any other person in Great Britain, is firmly of the opinion that steam cars are suitable as an experimental hobby only. He thinks they are unlikely

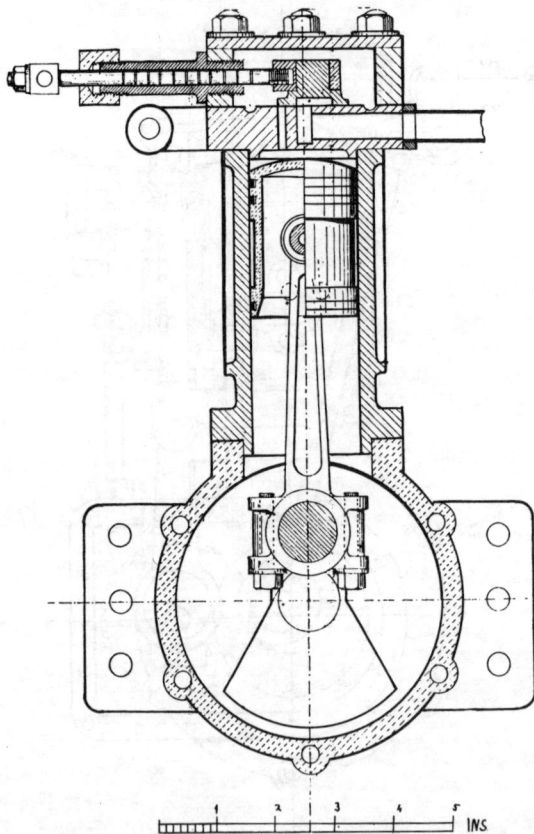

Fig. 3.8   Sectional side view of the same engine

to prove a practical proposition from a commercial angle. This view is shared by Mr. Kendrew, the constructor of the car mentioned above.

To those enthusiasts so inclined, a steam car project could provide a great deal of pleasure both in the initial construction and the subsequent development. Readers who seriously contemplate taking up steam car construction would be well advised to contact The Steam Car Club of Great Britain, which is an organisation having much interest in steam-driven road vehicles. Information and material is available to members and a quarterly magazine is published. The hon. secretary at the time of writing is Mrs. Diana Goddard, 29 Nibs Heath, Montford Bridge, Shrewsbury, SK4 1HL, United Kingdom.

CHAPTER FOUR

APPLICATION TO MODELS

It is known that Herbert Teague and V. W. Delves-Broughton applied a flash boiler to a model racing boat around 1908. The idea was borrowed from a steam aeroplane plant by Langley in the U.S.A. The new system with its high power to weight ratio very soon proved its value by establishing new speed records. The speeds attained were around the 20 *mph* mark—about three times faster than anything achieved before.

A prolific writer on the subject of the 'new steam' and model hydroplanes at this time was our old friend 'The Carpenter's Mate' (W. Blaney), and Messrs. Teague and Delves-Broughton freely acknowledged their indebtedness to Bill Blaney, 'but for whom *Folly* would never have been built.'

One development arising from the increased speeds was the introduction of the circular course for racing: it was soon found that stopping boats running free at 20 *mph* was hazardous in the extreme! The credit for this idea must go to Teague and Delves-Broughton, since no earlier references can be found for its use.

Since those early days, of course, it has been used for model cars, model aircraft and even full-sized aircraft—the last as a method of take-off in a confined area.

It was speedily discovered by exponents wishing to emulate the performances of *Folly* and *Incubus,* that that although there was no great difficulty in making a flash plant work, considerable problems occurred if very high performance was required. The number of exponents capable of fully exploiting the principle were comparatively small, although there were undoubtedly many who built flash plants with somewhat less success. Among the early pioneers may be quoted G. D. Noble, F. W. Westmoreland and A. Norman Thompson.

One of the most famous model flash steamers was Stan Clifford's *Chatterbox,* a craft still quoted with bated breath by some older enthusiasts. This boat set up a record of 43 *mph* in the early 'twenties which remained unbeaten for something like 15 years. With the state of development of petrol-engined craft in the late 'twenties and early 'thirties, it appeared almost unbeatable, and it was not until 1936 that it was officially exceeded.

. On of the attractions of the flash steamer was the fact that unlimited supplies of water were available if a simple scoop was fitted. This may well be the reason why flash steam has always been far more popular for model boats than for any other branch of model making.

The rules for the *Model Engineer* Speed Boat Competition at one time demanded that all steam-driven craft should carry their own water. This caused much heart burning among the flash steam merchants of the day, but there is now some evidence that for racing, it may not be a disadvantage at all!

In the case of tamer flash plants required to run for long periods, it is of course desirable to pick up pond water. The slower speed means less aeration of the water, and filters can be of adequate size to cope with the solid matter as weight is of little importance.

One of the most impressive examples of flash steam applied to free-running craft was 'J. Vine's *Silver Jubilee,* which ran regularly in regattas prior to the 1939–45 war. This craft was notable, not only for its straight running abilities, but also for the remarkable control of performance which could be achieved by merely adjusting the lamp flame. A well-known flash steam enthusiast has referred to the fact that on one occasion the boat was held with the engine just running quietly for at least 5 minutes then on opening up the lamp, away she went at about 10 *mph.* The water supply must have been exactly correct in relation to engine demand in order to behave thus. Incidentally, although the plant was fairly heavily constructed, and installed in a hull over 5 feet in length the maximum speed approached a genuine 20 *mph.* We well remember witnessing what must have been one of the last '100-yards straight' events ever held, at a South London regatta when *Silver Jubilee* won very easily.

In more recent times the number of enthusiasts employing flash steam has fluctuated at different times. In an early post-war regatta organised by the Model Power Boat Association flash steamers won every event. No less than eight were entered; at other periods the number of active members making entry at regattas has been reduced to one or two only. At the time of writing there are about half a dozen active in hydroplane racing and a similar number in other events. There are, of course, many who are very interested yet take no active part, and others who have boats on the shelf but hope to get going in the future.

Jim Bamford has demonstrated that it is quite feasible to develop a flash plant on the bench by employing a simple type of water brake (a propeller in a tin of water) and this assists to answer the criticism that a flash steamer 'takes too long to develop' or that a pond on the doorstep is a prime necessity!

For certain types of power-driven ship models, the flash plant has very great advantages in the way of low centre of gravity and low headroom required. *Plate 7* illustrates H. W. Saunders' steam yacht

Plate 7   The steam yacht *Verulam,* by H. W. Saunders

Verulam where the depth below decks is only 4 inches. The flash plant employed in this craft will be referred to later but it must be stated that this is a very successful boat that will run consistantly for over 30 minutes on one filling of fuel.

Model destroyers are nearly always constructed with draught far in excess of true scale in order to accommodate the plant. That flash steam could solve this problem was demonstrated by the article written in 1936 by 'W. F. W.' describing the model destroyer *Amazon*. The scale was $\frac{1}{66}$ of full size, giving a length of 55·6 inches, beam 5·73 inches and 1·6-inch draught, and in construction these dimensions were very nearly achieved, the draught being increased only a very little. The boiler used was 9 feet of $\frac{3}{16}$-inch plus 1$\frac{1}{4}$ feet of $\frac{1}{4}$-inch copper tube, engine was the Stuart 'Star' a good lightweight engine not now available. Firing was by a vaporising burner of L.B.S.C. pattern placed

Casing 3.3" dia. × 8" long

To engine

Exhaust pipe

1/4"

Top of exhaust pipe

15" figure 8" coils

5/16" tube wound
on 3/4" mandrel

From
pump

Asbestos

Oil trap
Oil outlet

Baffle plate

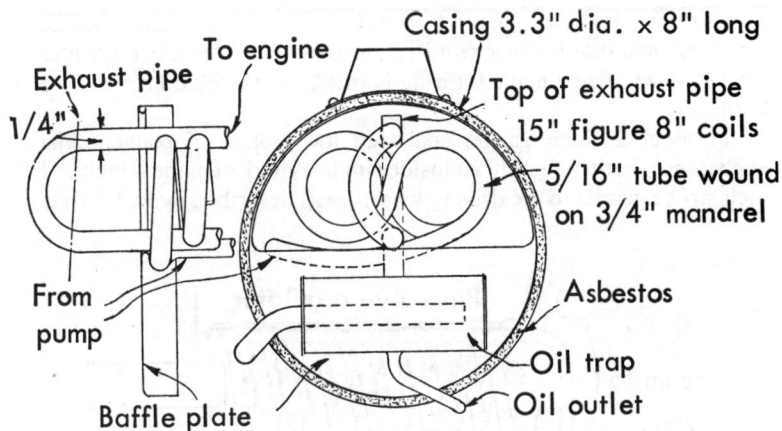

Fig. 4.1   The smoke box of the *Amazon* boiler, showing the
arrangement of coils

below the tubing. A cross-section of the boiler casing at the smoke box
is shown in *Fig. 4.1*.

Model locomotives

Although few model locos operating on flash steam appear to have
been built by amateurs, commercially produced locos were manufactured in large numbers by the firm of Carson's before the 1914–18
war. The principal output appears to have been gauge 1 L. & N. W.
'Experiment' type. The boiler arrangement is shown in *Fig. 4.2*,
consisting of a single coil lagged with asbestos and a vaporising type of
spirit lamp which blew a flame into the boiler coils. No feed pump was
fitted, the water being carried in a tank in the tender pressurised by a
cycle pump.

The late James C. Crebbin—known as 'Uncle Jim' to the model
engineering fraternity was at this time chairman of Carson's, and at a
later date was quoted as saying that the loco would haul an extraordinary load. He believed that it was rarely necessary to pump the water
container to more than 30 *psi*. A check valve between water tank and
boiler was found necessary to avoid surging.

Some of Carson's other engines had more elaborate flash boilers and
a geared feed pump arranged on the second tender axle. An engine built
to special order of Sir Harry Lopes was a ¾-inch scale *Caledonian*.

This locomotive was declared by Mr. Carson to be the fastest he had ever seen, and that it never required more than half throttle no matter what load was being hauled. Firing in this case was by a Primus type burner.

A model $2\frac{1}{2}$-inch gauge, coal-fired locomotive of conventional appearance, but using only an inside single-cylinder engine $\frac{7}{8}$ inch × 1 inch stroke geared to the driving wheels, was described by K. L. Myer

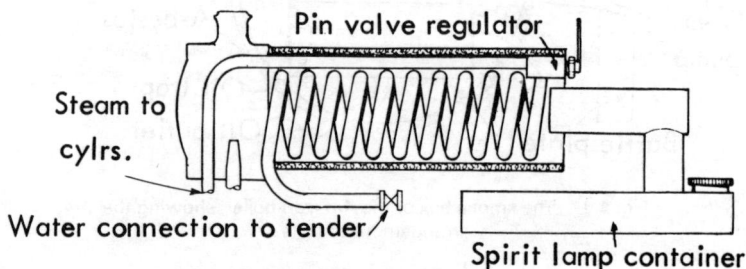

Fig. 4.2   A flash steam boiler for model locomotives, as described by Mr. J. Crebbin

in 1935. Although a full description of the flash boiler was given, including details of various modifications, it was eventually concluded that the rather poor results were caused by the boiler being too susceptible to slight variations in the burning of the coal fire. Feed water was by a crosshead driven pump. Since a 'Smithies' type water-tube boiler was also unsuccessful, it appears possible that part of the trouble may have been the single cylinder arrangement fitted with a single eccentric without any locomotive gear capable of 'notching up'.

The boiler was arranged in a similar way to the Carson engine. The fire door was sited at the upper part of the 'backplate', and although in an unorthodox position was found convenient for firing a tank locomotive. A total 25 feet of $\frac{3}{16}$-inch and 5 feet of $\frac{1}{4}$-inch copper tube of 20 g. were used in the construction together with shredded asbestos, *Pyruma* fire cement and sheet steel. A more successful effort was that by N. Dewhirst. In this instance, the loco was a $2\frac{1}{2}$-inch gauge 4–6–0 job with outside cylinders of $\frac{5}{8}$-inch bore and 1 inch stroke. Outwardly the appearance was conventional, and the fact that a flash boiler was fitted could not be seen without detailed inspection.

The boiler consisted of 17 feet of $\frac{3}{16}$-inch diameter copper tube arranged in a very similar manner to the original Carson engines and that of K. L. Myer (*Fig. 4.3*). No mechanical pump was employed but a hand pump was fitted to replenish water in the reserve tank, initially pressurised by separate air tank. The latter was pumped up by a cycle pump and usually operated at 50 *psi* to commence with, extra water could be injected by the hand pump to increase the pressure to 90 *psi*. Mr. Dewhirst stated that the boiler was not difficult to handle and the

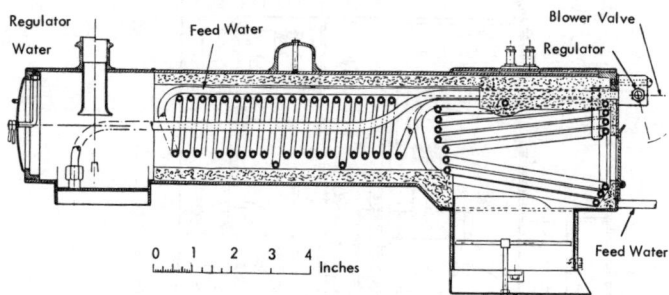

Fig. 4.3   The longitudinal section of a coal-fired flash boiler loco, by N. Dewhirst

fire easy to manage. At first, the tendency was to run the fire too fiercely. It was also stated that coal and water consumption was very light. Although no claim was made as to exceptional hauling power, 20 stone could be pulled using a single trolley.

A most interesting 2½-inch gauge *Sentinal* Railcar model was made by J. H. Rodgers of Leicester and subsequently described by 'L.B.S.C.' *Model Engineer* (see Appendix). *Fig. 4.4* shows cross-sectional details and several ingenious features emerge. The flash boiler was coal fired, and the firing chute fitted with a draughtdoor controlled by a simple thermostat (riveted bi-metal strip) attached to the steam manifold. When the steam reached high temperature, the draught door opened thus reducing the intensity of the fire. A single-cylinder, double-acting engine was arranged vertically with a large diameter flywheel, bevel-geared in a horizontal position. The bevel gear to the driving wheels allowed the engine to make three revs to each turn of the drivers; the latter being coupled by outside coupling rods.

Of particular note is the water feed arrangement. An axle-driven feed pump fed water to a closed pressure tank fitted with a pressure gauge and relief valve to return unwanted water to the water tank. Normal working pressure was 80 *psi*. The engine would run, hauling a heavy load of eight wagons for as long as fuel and water was supplied.

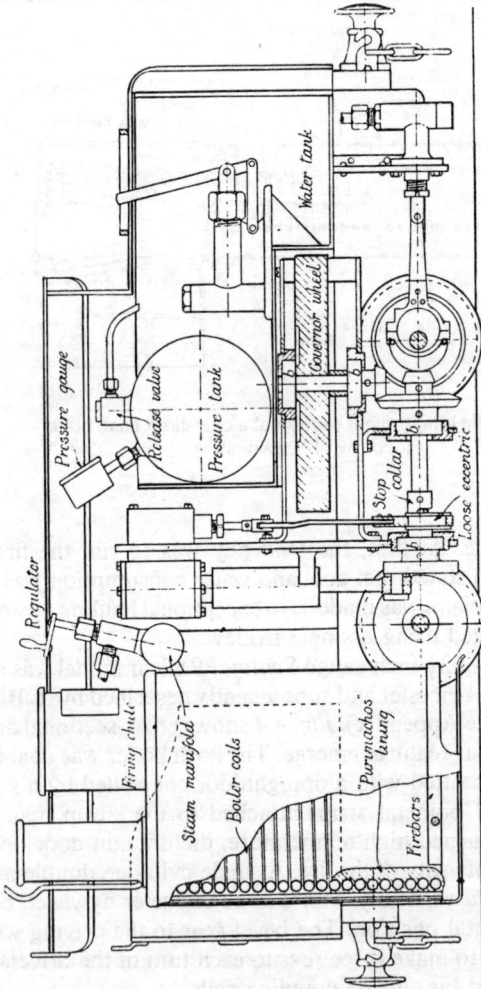

Fig. 4.4   Diagram of the 2½-inch gauge *Sentinel* loco by J. H. Rodgers

Lubrication to the steam chest was by oil pump fitted with a slight feed drip to the intake so that delivery was made only when enough oil was available. As the oil pump was driven by lever from the feed pump strap this arrangement avoided a large reduction gearing.

The last examples are surely pointers to the path of success with passenger hauling jobs. It appears that the uncontrolled flash plant is not feasible where varying loads and gradients are concerned. In particular, small passenger-hauling locos have similar problems in this direction as full-size locomotives and other steam transport.

As far as can be ascertained no model traction engines, road locomotives or steam cars with flash boilers have yet been built. Yet there are endless possibilities for experiment in this field, as well as for rail locomotives. A point which has been made on many occasions concerns the relative ease of making a flash boiler compared with the conventional patterns. This is particularly so when the task of constructing a typical coal-fired loco boiler is considered. Not only is the material now very expensive, but the construction and brazing of the larger sizes becomes a formidable task even to model engineers of experience. Many a model loco project has foundered on the boiler. rather than the lathe and fitting work.

Model aircraft

In the very early days of model aircraft, flash steam plants were seriously considered as a possible method of propulsion together with compressed air or rubber drive. The advent of model petrol engines small enough for aircraft work—largely pioneered by Edgar T. Westbury, marked a reduction of interest, which ceased completely when the small mass-produced ic engines came on the market.

Long before this period, H. H. Groves had produced flash steam plants for model aircraft propulsion. Mr. Groves was also a pioneer of flash steam hydroplanes and describing a visit to him, F. W. Westmoreland wrote in 1916, "On Sunday, Sept 26th, we decided to get out early in the morning with a large and beautifully made steam-driven monoplane. Here I found plenty to learn but, unfortunately no free rides. After making several attempts to outdo 'Fokker' speed the machine struck a bank and came to grief breaking the chassis in two. Bang went many hours of patience and labour, only known to model makers."

In 1936 an account of one of the latest 'Groves' aircraft plants was published in the *Model Engineer*. This must surely have been one of the lightest model aircraft flash plants ever constructed. Engine, pumps, boiler, lamp and water tank weighed but 5 ozs.

The whole plant was mounted on a chassis of aluminium beaten into a channel section. Length from tip of the propeller shaft to the end of the boiler was $8\frac{1}{2}$ inches, width was about $3\frac{3}{8}$ inches.

Interesting features of the engine included the inverted position and the use of a slide crank to reduce overall height (*Fig. 4.5*). Bore was $\frac{1}{4}$-inch by $\frac{7}{16}$-inch stroke, double-acting. Piston valve $\frac{1}{8}$ inch diameter by $\frac{1}{8}$-inch stroke, driven by follower crank from the main crankpin.

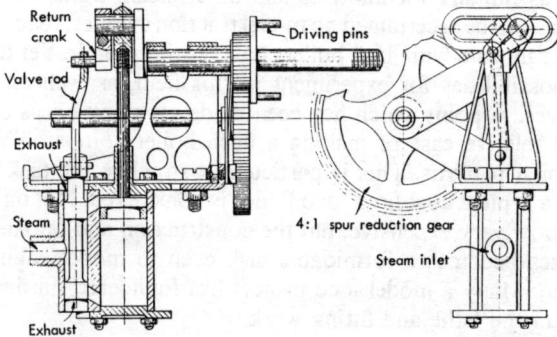

Fig. 4.5   The double-acting engine by H. H. Groves

The boiler was a coil of $\frac{1}{8}$-inch diameter stainless steel tube enclosed in a casing of ·005-inch shim steel backed with $\frac{1}{32}$-inch asbestos; holes $\frac{7}{32}$-inch diameter pitched $\frac{1}{2}$ inch apart perforated the casing which had the end closed. Feed pump was $\frac{1}{8}$-inch bore, the stroke variable from $\frac{3}{16}$ of an inch to $\frac{1}{2}$ inch. The water tank measured $1\frac{1}{4}$ inch diameter by $2\frac{1}{8}$ inches

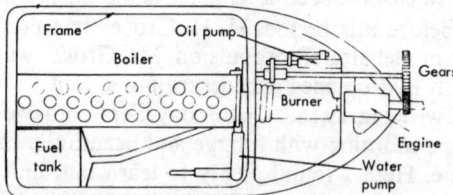

Fig. 4.6   The layout of the model aircraft flash plant designed by H. H. Groves

long. *Fig. 4.6* is a diagram of the plant arrangement in plan view. The water tank is not depicted as it was sited within the cowling.

Many years after the construction of this plant, a replica of the boiler and lamp was made by Prof. D. H. Chaddock, when experimenting with small steam turbines (*Plate 8*). He reported that it operated precisely as H. H. Groves stated. Evaporation on a test bench being about 4 ozs. of water per minute, was slightly better than the original. The perforated boiler casing allowed intake and exit of air as required and combustion was good. Burner performance was excellent and if

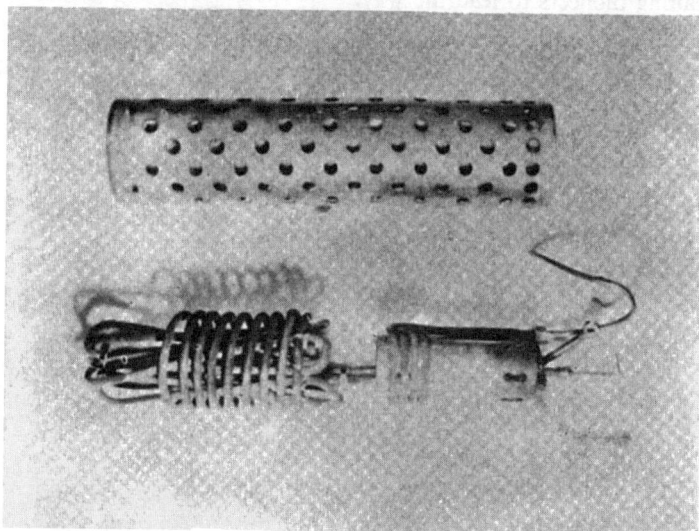

Plate 8   Prof. Chaddock's replica of the 'Groves' flash boiler and lamp

turned up high without water the coils soon reached a yellow heat.

Very recently a larger replica of this boiler was used to power a flash steam 'mini-bike' constructed by Richard J. Smith in the U.S.A. and described by J. Hermans.

It must be conceded that the application of flash steam to model aircraft at the present time is most unlikely. Apart from the fact that few past constructors have been able to produce very lightweight plants,, the safety aspect of flaming blowlamps in free flight becomes somewhat alarming! There exist possibilities in the way of control line flight and, of course, radio controlled models, but it would need an exceptional enthusiast, with the necessary experience in several different fields as far as the latter is concerned.

From the foregoing, conclusions may be drawn that model boats and hydroplanes offer the least difficulty and the most popular field for experiment. There have been many craft fitted with flash steams plants over a long period of years and some information has been published.

In the later chapters of this work we have endeavoured to incorporate the most important and basic matters together with latest practice.

As far as other branches of modelling are concerned the field is wide open, and flash steam locos, traction engines, etc. are really still awaiting pioneers to lead the way.

CHAPTER FIVE

MODEL FLASH BOILERS

The power of any steam plant lies initially in the combustion of the fuel and the efficiency of the boiler; the engine is simply a device to change the energy contained in the steam into mechanical work. For this basic reason it is impossible to expect high output from a plant unless a fair amount of fuel is consumed and *usefully* employed in heating the boiler tubing.

So we are dealing with boilers and firing methods together, as they are complementary.

We are mainly concerned here with high performance boilers. Although the general details can be applied to plants of lower power, the problems in the latter type are far easier to overcome. Less fuel is required to be burnt, thus less air is required and boiler casings can be heavier with more efficient lagging.

Materials

There is no doubt at all as to the best material—it is stainless steel! Unfortunately, small bore, solid-drawn tube is rather expensive at the moment—perhaps because of low demand when compared with the larger bore continuously welded tube now used for many engineering purposes.

In spite of the expense factor it must be remembered that a stainless steel boiler will outlast all others. Ordinary mild steel tubing usually lasts about two seasons if applied to regular racing work in a hydroplane. Stainless boilers are known to have lasted at least ten years and are still going strong. One of the few snags with stainless is the fact that in small sizes it seems difficult to obtain long lengths—about 10 feet appears to be the maximum. This means joints, and although sif-bronze will last for a while it usually becomes porous after a season or so, unless well protected from the direct heat of the flame. The welding of thin stainless tubing to make a non-porous joint capable of with-

55

standing high pressure without occluding the bore is a job for an expert welder used to welding thin stainless steel. If welding is preferred professional assistance must be obtained.

The second choice of material is mild steel and indeed many flash boilers have been constructed of this metal both in model and full-size practice.

Most M.S. tubing of the smaller sizes may be readily coiled cold without annealing. If not, heating to red heat in sections by means of a blow lamp will provide adequate softening.

A readily available type of tubing is that sold by motor factors and intended for hydraulic pipe lines on vehicles. Care must be taken here; some of this tube is not seamless but rolled up in a special manner and copper-brazed. Another problem is the protective coating on the outside which appears to be some form of 'tinning'. One sample of this material showed severe pitting after a short period of use as a blowlamp vaporiser. Replacement with a fresh coil that had had the coating scraped off gave no trouble.

Copper tubing is probably the most readily obtainable material for flash boilers, and a fair number of successful cruising boats have employed this metal for the boiler coil. The advantages include easier coiling up and re-coiling, if necessary! Any scale formed is soft, causing little damage to the engine and the heat conductivity is better.

An example of a boiler employing *brass* tubing is shown in *Plate 9*. This is the boiler of the steam yacht *Verulam* by H. W. Saunders of the St. Albans S.M.E. The boiler coil is made from $\frac{3}{16}$ inch-diameter × $\frac{1}{8}$-inch bore brass tube as used for 'Tecalemit' pressure grease lines. The total length of the tubing is about 12 feet. This boiler has been in use several years and shows no signs of scaling or wasting.

While adequate for moderate power plants the use of copper or brass in a racing hydroplane is rather dubious. The melting point of copper is about 1,000 degrees C, and providing that the water supply is sufficient to keep the coils below a dull red heat, all should be well. It must be remembered that the tensile strength of copper is much reduced at high temperature, and the risk of tube failure would be greater in a high pressure plant.

For example, it is known that engines fitted with poppet valves giving short cut off tend to operate on high pressures (over 3,000 *psi* in one of J. Bamford's engines). Should the engine stop altogether it is almost certain that under these conditions the boiler would burst or melt! If, of course, the firing is by fuel pump and not by pressure tank, the flame dies automatically and this is a tremendous advantage in any boiler. There have been attempts to incorporate safety devices in flash steam hydroplanes using pressurised lamps. One such device was made by H. Turpin. A hydrostatic valve released the container pres-

Plate 9   The flash boiler for the steam yacht *Verulam* by H. W. Saunders

sure when the water pressure dropped (*Fig. 5.1*). The device was not entirely reliable, probably because there is often a pressure build-up if the engine stops with water still in the boiler.

An alternative might be to use a mechanical device operating on

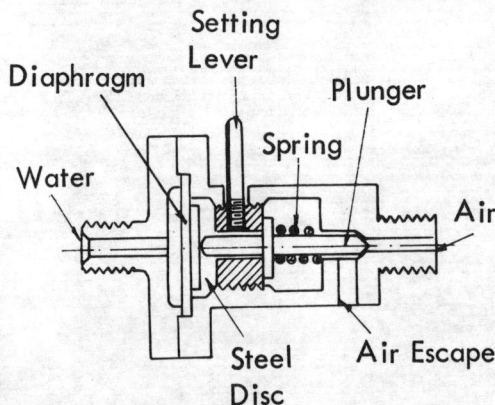

Fig. 5.1   Hydrostatic air release (Turpin)

similar lines to the old-fashioned spring gramophone governor geared down from the engine. The governor could be made to trip a release fitting when the engine revs dropped too low.

Size and lengths of tubing

By far the most popular size of tubing has been $\frac{1}{4}$-inch outside diameter and around 20–22 S.W.G. wall thickness. Some constructors have used some $\frac{3}{16}$-inch O.D., and some $\frac{5}{16}$-inch O.D. tube, but few have made boilers either entirely of $\frac{3}{16}$-inch, or entirely of $\frac{5}{16}$-inch diameter. There has been a small but persistent school of thought that maintains that smaller diameter boilers are the more efficient, and this view was supported by an article that appeared in *M.E.* July 1931. The writer, 'E. T.' gave details of tests carried out on three copper boilers each 11 feet in length and $\frac{3}{16}$-inch, $\frac{1}{4}$-inch and $\frac{5}{16}$-inch diameters, respectively, and wall thickness ·03 of an inch. Each coil was wound in a tapered form, $2\frac{1}{4}$ inches inside diameter down to $1\frac{1}{2}$ inches diameter and spaced $\frac{1}{8}$ of an inch apart. The casing left a $\frac{3}{8}$-inch gap at the largest coil. Water was fed from a water pump driven from an electric motor.

Each boiler was fired from the same air-gas blow pipe $1\frac{7}{8}$ inches in diameter, and various evaporation tests conducted with a spring-loaded outlet valve set to blow at 500 *psi*.

In each test the $\frac{3}{16}$ inch diameter boiler gave the best results, and on a maximum evaporation test managed 27 cubic inches per minute with the gas blowpipe flat out and the steam highly superheated. This represents about 1 lb. per minute and seems a remarkable figure for only 11 feet of tube and the fact that this boiler had the lowest heating surface. In every test the $\frac{5}{16}$-inch boiler gave the worst results. On repeating the experiments using thicker walled tube the relative results were confirmed but evaporation increased by about 12 per cent.

It should be remembered that bench tests on flash boilers under sheltered conditions may give very different answers from actual results on the pond. The water and steam velocities have to be very high if the figure of 1 lb. per minute is to be achieved with the $\frac{3}{16}$-inch tube and there is bound to be a larger pressure difference between the water and steam ends hence the water pump would have to do more work. Where the pump is driven by an independent souce of power, as in the tests referred to above, the additional work required is of no importance. However, it might be a different story when the poor old engine has to do the work! It is quite possible that the $\frac{1}{4}$-inch O.D. size represents the best comprise between boiler efficiency on the one hand and driving pumps on the other. What can be said is that many enthusiasts have tried small bore tubing for racing boats without significant results. It is true that one *M.E.* reader quoted a $\frac{5}{32}$-inch diameter boiler as one of his best, but it is thought that it powered a non-racing job.

As much as 60 feet of tubing has been crammed into an A-class flash steamer. This was B. Pilliner's *Ginger* (1947/50). This length was relatively inefficient and a decided improvement occurred when it was reduced to 42 feet. The later boilers were only 33 feet of $\frac{1}{4}$-inch O.D. tube, yet the later boat *Eega Beeva* officially achieved 59 *mph*, and over 70 *mph* in practice. It was on the 60 foot boiler that a pressure drop of no less than 400 *psi* was recorded; the higher pressure naturally being at the water end.

Boiler casings

The traditional casing used for flash boilers in model boats consisted of a rectangular or circular casing open at the blowlamp end and having an uptake at the other. Sometimes additional louvres were cut in the casing to assist combustion. As always the problems arose when racing boats were involved. The larger amount of fuel to be burnt, the motion of the boat, stray steam, etc., all made a difficult task of keeping the boilers hot or even the lamp stay alight. The main failing of the old-time flash steamer was in keeping going—one or two brilliant laps were usually followed by slowing down or petering out.

A course of action often recommended was to run the boat at dusk—the idea being that it was easier to see if combustion was in order. The idea worked all right—usually the flames appeared everywhere but inside the casing!

This particular problem was frequently blamed on the torch-pattern blowlamp. E. T. Westbury suggested that experiments should be carried out with diffused flame burners, pointing out that it was better practice in that a greater number of small, less intense flames should be used rather than one or two large roaring blowlamps. *Fig. 5.2* shows a

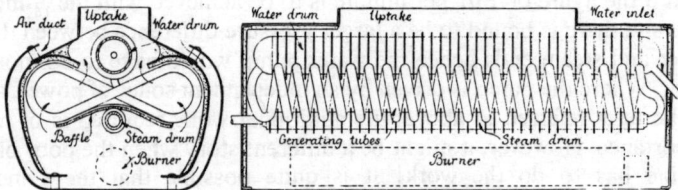

Fig. 5.2   A boiler with 'figure-of-8' tube arrangements and
steam and water drums

suggested design by E. T. W., but it is not known if any examples have ever been constructed.

Many of the early steam cars employed a circular boiler casing with the tube arranged in flat coils and sometimes having a type of diffused flame burner underneath. Others such as the 'Doble' had the combustion space at the top and combustion occurred above atmospheric pressure, since all the air required was supplied by a blower.

A design by Thomas Hindle for a model boiler on similar lines, which has never before been published, is shown in *Figs 5.3* and *5.4*. It will be noticed that the fuel is fed by a carburettor of the slot-diffuser pattern using a wick-type, spirit burner for ignition purposes. Exhaust steam from the engine passes through a blast pipe and nozzle, and the draught created draws air into the boiler via the venturi. Combustion occurs in the annulus immediately above the flat coils, the hot gases passing through them before passing to the central flue. No doubt some experiment with nozzle sizes would be required but this should present no great difficulty. The provision of a steam scale trap (not shown) of the centrifugal pattern is interesting, and the whole design offers much food for thought. The likeness to steam car practice is probably intentional, as Mr. Hindle is well-known as an expert of many years standing on steam cars and experimental steam and steam engines.

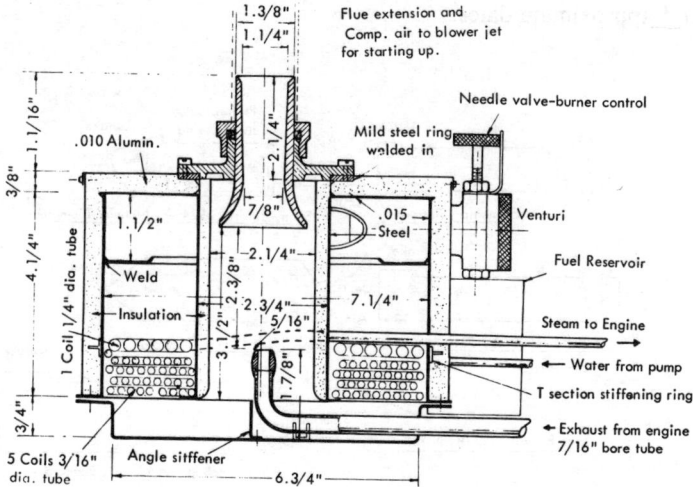

Fig. 5.3   The design by Thos. Hindle showing section of the monotube steam generator

The early users of conventional horizontal boilers fully realised that 'keeping the boiler hot' was the major problem in flash steam hydroplanes, but the open-ended boiler casing remained in use for over 30 years except that baffles and shields tended to be used more frequently. In one case (*Ifit 6*) bending one of these baffles $\frac{1}{8}$ of an inch gave a 5

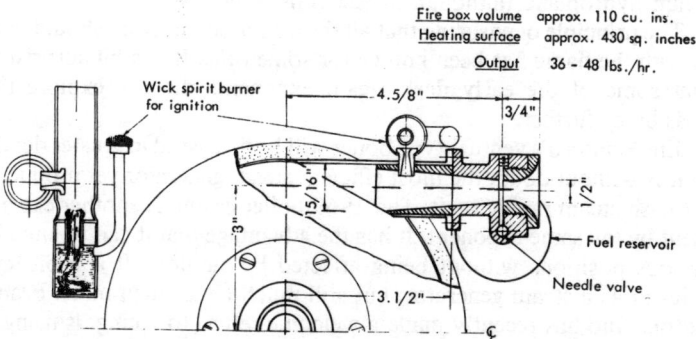

Fig. 5.4   The burner and ignition system of the Thos. Hindle generator

mph increase in speed. Torch-pattern blowlamps also ruled the roost, the only modification being in the number of burners used. Twin burners were mostly favoured especially for A-class hydroplanes. *Fig. 5.5* is intended to show the pattern of development, with constructors and approximate dates.

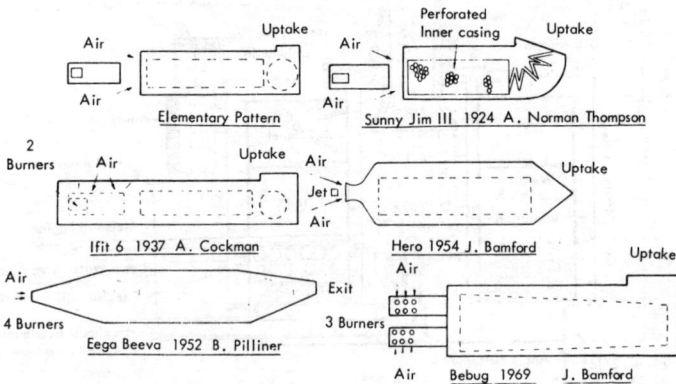

Fig. 5.5   The development of boiler casings and style of burners between 1924 and 1969

It will be noticed that the first exponent of the completely enclosed boiler was Bernard Pilliner. The forward air scoop required some adjustment in size but when the correct balance was found the effects were dramatic. Indeed, he was the first to exceed 60 *mph* with a flash-steam hydroplane (although in test runs only).

The principle of ensuring that all the air for combustion should pass in with the flame has been known for some time. It is a bit surprising that some of the early flash steam exponents did not explore the possibility further.

Jim Bamford's venturi-type boiler will be discussed in greater detail, as it is without doubt the most efficient steam generator yet produced for flash steam racing craft. The 1969 boiler casing is another experiment by the same exponent. It has the advantage that it can be sited in various positions without being affected by air flow. It is a bit less efficient as a steam generator, but still much better than most. Frank Jutton, who has recently made a welcome return to racing, is using a Bamford 'Mark I' type with great success. The speed attained is around 20 per cent greater than with the older boiler.

Contra flow and types of coil

In theory, water should be fed to the coolest end of the boiler, the temperature difference ensuring that the water gains heat. If the water is fed to the hot end, the cooler part of the boiler does little or nothing, as the steam may be already of equal temperature and thus can gain no further heat.

In practice, there are many examples known where this rule has been flouted! To be fair, this arrangement of feeding to the hotter coils was sometimes done deliberately in order to protect the tubing especially if copper was the metal used. Even mild steel tube can scale rather badly, since the first few turns of a flash boiler can get very hot indeed in the direct flame of a powerful blowlamp. Sometimes the coils were arranged in such a fashion that the section exposed to the direct blast of the lamp was fed with water. Then the tube was taken to the cooler end. It must also be admitted that certain exponents have tried feeding to the cool part, then, as an experiment, reversed this to the hot end without noticeable difference. This, of course, does not necessarily invalidate the principle, things are not always what they seem—especially in flash steamers!

Plate 10   The flash boiler coils of Eega Beeva (designed by
B. Pilliner), with the upper casing removed

The simplest flash boiler consists of tubing wound in a cylindrical coil with the flame of a 'torch' type blowlamp played through the centre. There is some tendency for the flame to shoot through the whole coil and the most common expedient was to fit a cross coil at the end which served to baffle the flame. On occasion this cross coil glowed bright red while the rest of the boiler appeared cooler, but this was usually when steam was taken from this end.

Critics of the tightly coiled flash boiler have often suggested that each coil should be well spaced from its neighbour so that the heating surface was effectively exposed to the flames. A good example of this is the boiler of *Eega Beeva* shown in *Plate 10*. It will be noticed that the coils are spaced around at a distance of half an inch apart. Another example where the tubing was well exposed to the flame was the boiler designed by H. H. Groves for his model aircraft plant. On this model a number of longitudinal zig-zags were sited in the centre, thus splaying the flame outwards between the loops of the main coil.

Although it might appear that direct exposure of tubing *ought* to be the more effective, there is no clear evidence on this point. It is more likely that *effective combustion of the fuel within the boiler casing* is the vital consideration, irrespective of the actual arrangement of the tubing.

Flash steam enthusiasts have long been aware of the problem of pressure drop from water to steam ends. In an effort to reduce it they have tried parallel lengths of tube. Sad to relate such experiments have usually failed, one side getting red hot while the other remained black. This shows that only one side had water flowing. A successful example was demonstrated by A. Cockman's *Ifit VII*, in which part of the boiler tube was in parallel. An interesting experiment would be to have parallel boilers fed by separate pumps, connected at the steam end only. However, the additional complication of another pump might easily offset any gain. As far as is known this particular experiment has not yet been tried—at least on model hydroplanes.

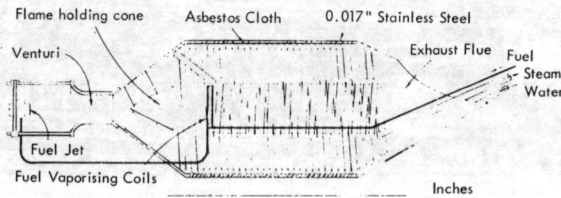

Fig. 5.6   The venturi boiler designed by J. Bamford

The 'Bamford' venturi-type boiler is undoubtedly the most efficient steam generator yet devised for model hydroplane work. *Fig. 5.6* shows the layout. It is really a blowlamp and boiler combined, and all air necessary for combustion passes through the venturi. Forward motion of the boat assists the natural injector effect. In Frank Jutton's version of this boiler combustion improves considerably as the boat builds up speed.

The casing of the original was about 6 inches diameter, and total overall length 22 inches. As can be seen, the contra-flow principle is followed in the arrangement of the centre coil, but not in the outer one. Mild steel tubing of $\frac{1}{4}$-inch diameter was used in the original boiler, but a number of failures occurred due partly to the high pressures attained. About 2,000 *psi* was the normal working pressure on an experimental turbine plant, and even higher values were recorded with a piston valve engine having the cut off at 65 degrees after T.D.C. About 30 feet of stainless steel tube was used in the final version of the boiler, which is the one illustrated. An earlier type shown in *Plate 11*, installed in *Hero*,

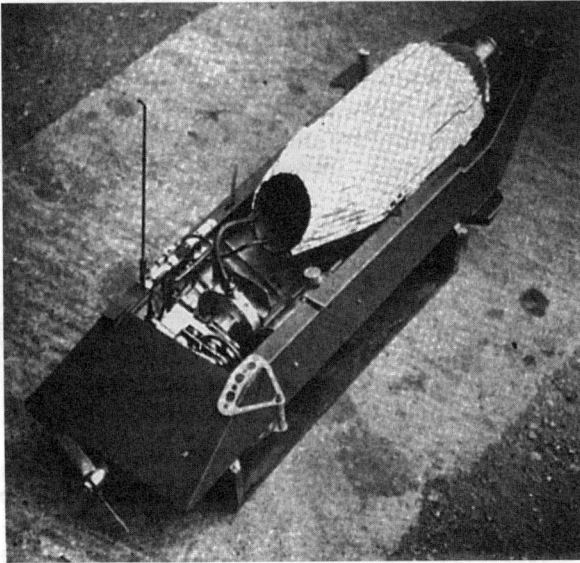

Plate 11   The 'Bamford' venturi-type boiler in its turbine-driven hydroplane. The exterior lagging is aluminium foil

was lagged on the exterior with thin asbestos and aluminium foil. The boiler was efficient but made an extraordinary booming noise which could be heard over a considerable distance. It was thought that this phenomenon was due to the venturi and boiler acting as a kind of pulse jet. Fortunately the fault was cured in the later model. The noise is considerably less than now required by the M.P.B.A. for racing craft.

Single-jet versus multi-jet burners

The firing of the Bamford boiler utilises a type of vaporising lamp with the vaporising coils placed more or less directly in the flame and having a single large jet. The vaporiser tends to get rather hot and on one occasion the silver-soldered joint at the jet blew apart—presumably due to the effects of transmitted heat and pressure exerted by the fuel pump. It is strongly urged that all boilers made to this design should have a welded or at least sif-bronzed joint, whereby the jet block is attached to the vaporiser tube.

Fuel pumping presents problems—largely due to vapour locking and very low viscosity of the liquid. Jim Bamford at one time used 50/50 paraffin and petrol, but better reliability has been achieved with more paraffin and less petrol. Incidentally, the jet size was around ·030-inch diameter but it should be remembered that different operating pressures may require some adjustment. There seems no reason why an adjustable jet should not be tried, except that the flame pattern might be modified and fail to stabilise on the cone.

The 'Pilliner' boiler had quadruple burners of the dimensions shown in *Fig. 5.7*, and the vaporiser coil was also placed directly in the lamp

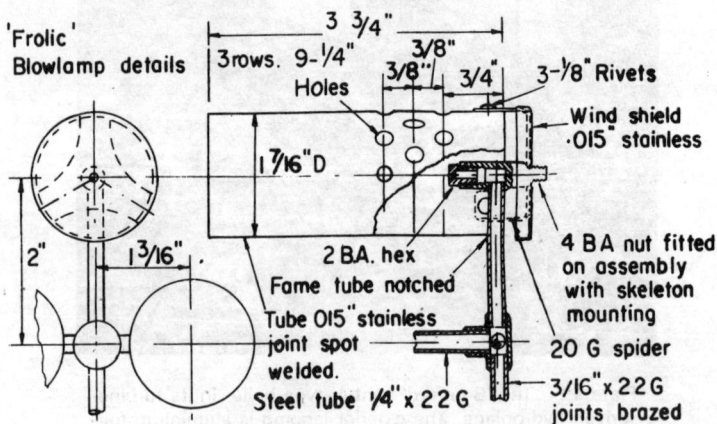

Fig. 5.7   Quadruple burners, designed by B. Pilliner

flame. It must be recognised that this boiler was also an extremely efficient steam generator. Although actual evaporation tests were not carried out, a calculation based on an estimated pump speed of about 2,800 strokes per minute and 90 per cent volumetric efficiency gave over 2 pints per minute. Even if the per cent efficiency was lower than this, it would still rank very high. Pump speed based on 2·8 : 1 gear ratio is thought to be on the conservative side, since it was assumed that propeller slip was only 25 per cent when travelling at 60 *mph*.

The best results were achieved when the optimum size of forward scoop was found by trial and error methods. When starting off incomplete combustion was evident by the yellowish appearance of the flames; on working up to speed and scooping up sufficient air combustion appeared excellent—at least as far as could be seen from the safe distance of the pondside.

A similar boiler employing 3 burners was constructed by A. Cockman for *Ifit 9, Plate 12*. A section of the casing hinged forward for

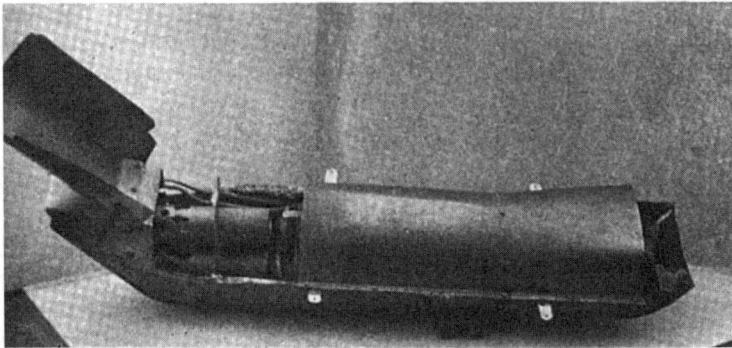

Plate 12   The boiler for *Ifit 9*, designed by A. Cockman,
showing hinged flap for starting lamps

starting up, and this was left open as the boat was released. Air pressure soon pushed it down to the 'closed' position which thus presented the air scoop correctly. Although no evaporation figures are available for this boiler, it must be classed as very successful as speeds approaching 60 *mph* have been achieved on many occasions.

In times gone by, criticism has been made concerning multi-burner blowlamps as being the rule of the 'big stick', but in retrospect it must be admitted that many of the single burner lamps then in use emitted excessively long flame if the jet was opened up. Flames 18 inches long were not easy to use in most boiler casings! The fact must be faced that if evaporation of 2 lbs. of water per minute is to be approached—and this is what is wanted—much fuel is required to be burnt.

It is now evident that a single-jet blowlamp can be made to burn with a shorter flame if the gas is efficiently mixed with enough air and perhaps a flame stabiliser provided. The combustion is complete

Plate 13　An experimental triple-coil boiler by F. Jutton
(engine and pumps on right)

within the casing, and the efficiency of the whole boiler becomes commensurably greater. Another factor in favour of the single jet is that modifications are simple when compared with a multiple-burner affair.

Feed pumps and pressures

We are somewhat hesitant about recommending the beginner to start out using fuel pumps. Jim Bamford has recorded that at one time he had great trouble in pumping fuel satisfactorily and made up a succession of pump valves of different materials without significant improvement. Finally, a new pump was made using ordinary ball valves, and it worked perfectly!

Our old friend 'Ifit', otherwise Arthur Cockman, also has had some mysterious happenings. On *Ifit 9* light pressure was used in the fuel tank in order to avoid a hand pump when starting up. The lamps commenced to vaporise and the engine started, but unless the tank pressure was then released the fuel supply stopped. One would imagine that the light pressure feed would assist rather than hinder. Needless to say, the pump valves were quite tight with restricted lift, and no reason could be found for the trouble.

It is very essential to keep fuel pumps cool and a water jacket using by-passed feed water might be best. Alternatively, the dual water/fuel pump as used by Jim Bamford has worked well—especially on paraffin/petrol mixtures.

We have conducted tests on a fuel pump of $\frac{1}{16}$-inch bore, discharging through a spring-loaded valve and connected to a pressure gauge. Using blowlamp fuel (a pure petrol) 80 *psi* was the maximum pressure attainable by hand pumping. Paraffin increased this to 120 *psi*.

It is not known how the relative efficiency of the pump compared with others but the experiment shows quite clearly that the pressures should be higher than the conventional fuel feed from a tank pressurised by a bicycle pump. The maximum pressure attainable by the latter seems to be around 60 *psi*—and, of course, the pressure is falling as the fuel is used up. Those racing craft successfully pumping fuel are very noticeably free from the fading out troubles that have plagued flash steamers in the past. Always providing the lamp remains alight!

Alternative burners

Mention has already been made of diffused flame burners and a design of E. T. Westbury's is shown in *Fig. 5.8*. This would be very suitable for plants of moderate output but the boiler coils would have to be

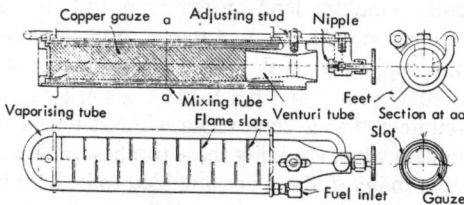

Fig. 5.8   A 'diffused flame' vaporising burner, suitable for firing water-tube boilers and flash boilers (of integral design)

arranged above the burner making for a higher boiler casing. This would not matter in many cases since a conventional water-tube boiler would be similar in height and considerably heavier.

In the past, a few water-tube boilers used methylated spirit as the heat source, but only two records can be found of any flash-steam boilers fitted to boats that have used spirit firing. It may be a possibility for a 'tame' plant but it must be stated that few simple burners seem to emit much heat and the vaporising pattern has a reputation for being unreliable.

A much more promising method of firing lies in the small propane or butane burners now freely available and sold in a variety of makes and different sizes.

A number of steam-driven prototype craft are at present using such burners very successfully for firing centre-flue and water tube boilers. There is no reason why they should not be equally suitable for flash boilers of moderate power. The great advantage of a gas burner is the fact that all you have to do is to turn on and light up! It also has the merit of cleanliness, burning with a smoke-free flame. The following information is based on Edgar T. Westbury's articles in *Model Engineer* during 1967 entitled *Bottled Gas for Model Boilers*.

Both butane and propane have calorific values not differing too widely from that of petrol or paraffin. The burners operate on the Bunsen principle, but employing internal baffles to ensure adequate mixing of air and gas. The 'Max Sievert' range are marketed by Wm. A. Myer Ltd., 9/11 Gleneldon Road, S.W.16, the sole concessionaires. Several burners are suitable for firing model flash boilers depending on the heat output required. A 12-ounce refillable gas container is thought to be still obtainable, which can be directly or indirectly attached to the burner.

The 'Veritas' butane gas burners have throwaway containers, which should not be removed from the burner until exhausted. 'Veritas' appliances are manufactured by Falks, 91 Farringdon Road, London. Other manufacturers include Ronson Products Ltd., of Leatherhead, Surrey, who have produced the Veraflame torch having a die-cast burner head. One appliance employing a throwaway container is the Bluet (Gaz) picnic stove and another is the Taymer torch which has a screw-fitting container that is removable without losing pressure.

The main problem when applying these torches to flash boiler work lies in the fact that many of them are intended to be held in the hand, with the burner presented at an angle relative to the container. This arrangement is usually unsuitable for model boats, and some modification may be required in order to site the fuel container in a more suitable position.

Certain burners, such as the Sievert Nos. 2950 and 2951 neck-tube burners, are made integral with a length of pipe of about 4 inch. This can be straightened out after annealing and connected to the refillable gas container. Burner heads Nos. 2940, 2941 and 3941 would also appear to be suitable. The first one listed is of steel and recommended for outdoor operation the flame offering good resistance to wind. The largest of those quoted here is the 2941 which gives a powerful flame which consumes 21 ounces of gas per hour.

A method of adapting the small flat type of gas container is depicted in *Fig. 5.9*. This affords far better clamping than the usual spring clips.

It should be emphasised that great care must be taken with any modifications to produce safe working. The gases used are heavier than air and can accumulate in the bilge of a boat. In full-size craft this can be hazardous and leakage of gas due to careless use or improperly fitted joints is the commonest cause of trouble. While no violent explosion is likely to occur in the hull of a model boat, some fire damage could ensue. Those already experienced in the nastier habits of petrol and paraffin burners will know all about singed hair and eyebrows!

No one as yet appears to have tried the use of bottled gas for firing a

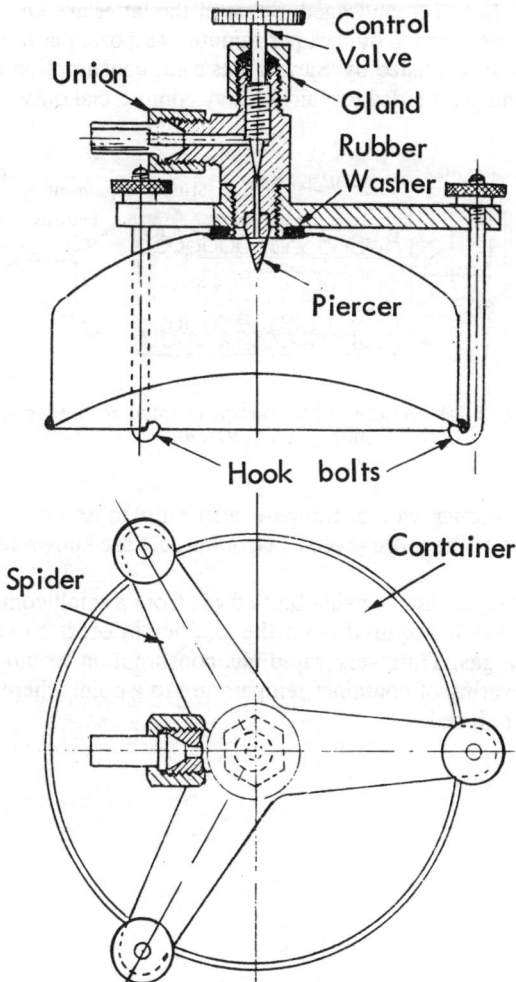

Fig. 5.9 Fittings for adapting a flat butane container
(Bleuet type) to flash boilers

racing boat. It might be possible using some of the larger burners
designed for brazing work, and this is one more field that is open to
experiment. A design by E. T. Westbury, for a flash boiler fired by
bottled gas is shown in *Fig. 5.10*, but this was not intended for high
output. A problem would probably arise if heat output equal to some

petrol blowlamps is envisaged. Some of the latter are known to have consumed well over 8 ounces per minute (480 ozs. per hour). One of the larger burners listed by 'Sievert' has a gas consumption of 145 ozs. per hour and is intended for fairly heavy commercial duty. In fact, this

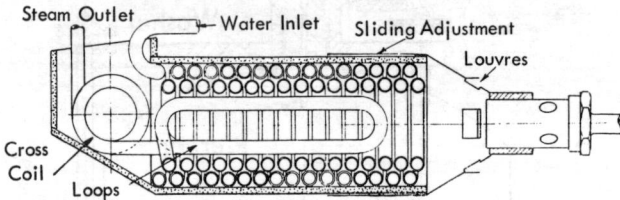

Fig. 5.10   A flash boiler design adapted to bottled gas fittings (E. T. Westbury)

particular burner would compare well with most boat blowlamps, since some of the more greedy petrol burners are known to have been inefficient.

One disadvantage of using bottled gas from a small container is the fact that heat is required from the exterior in order to vaporise the liquid to a gas. Thus very rapid gas consumption would result in a drastic lowering of container temperature to a point where gas output might be reduced.

CHAPTER SIX

ENGINES AND VALVE GEARS

Almost any model engine may be used for flash steam plants requiring only low performance. The exceptions are oscillating engines of the very simple pattern and those engines involving a soft-soldered construction, in which case the steam temperature may easily rise to cause the solder to melt and the engine to disintegrate!

For moderate and high performance work the chief factors concern materials and engine efficiency. Since efficiency is largely tied up with valve arrangements this will be discussed separately later in the chapter. With a model plant the complications of thermostatic devices controlling the steam temperature are usually out of the question, so the likely maximum steam temperature is somewhat uncertain.

Some materials are very definitely unusable as far as very hot steam is concerned, because of excessive wear of sliding parts. Thermal expansion must also be taken into account so that the combination of materials does not make the engine seize up when things get warm.

As a good general rule, for all working parts exposed to steam brass is best avoided altogether, and bronze is suitable only for moderate performance. For the out-and-out racing job, when it is not unknown for valve chests to glow a dull red, something better is required. The following materials have been used successfully:

Cylinders and valve chests	Cast iron (or meehanite) mild steel Low nickel steel, nickel chrome steel
Valves	As above, plus stainless steel
Pistons	As above, plus aluminium alloy

All these metals must be used in different working combinations with the exception of cast iron, which has the well-known property of working well as cast iron to cast iron. For piston valves, a successful combination is a cast iron or 'meehanite' valve working in a mild or alloy steel box to ensure that the valve does not seize up, the coefficient of linear expansion for iron being slightly less than that for the steel. Cast iron or 'meehanite' (trade name for an iron having a modified graphite structure) is usually one of the two metals used. The possibilities of case-hardening or nitriding of steel to resist wear have often been discussed, but there are problems. A study of the softening curve for high carbon steel (the material of the 'case') shows that at quite moderate temperatures the hardness will be reduced (tempering). At 300 degrees C, which is a full blue on a colour tempering chart the hardness values are naturally lowered.

Nitriding is better since the surface is harder and the hardness is not reduced significantly by moderate temperature. However, it is not suitable for the amateur since the job must be heated for many hours in an atmosphere of ammonia. One advantage is that no distortion occurs as the temperature of the process is only 500 degrees C, thus finished parts may be treated. Some firms have offered a service in the nitriding of car crankshafts as part of a 'hotting up' process when preparing sports cars for racing. An approach to one of these firms might be tried if this process is required. Some authorities state that ordinary steels cannot be nitrided properly and the process is only satisfactory if the component is made from 'nitralloy'—a steel containing aluminium, or other special alloy steel. Should aluminium alloy be used for pistons adequate clearance must be arranged to allow for the greater expansion of the piston compared with cylinder.

The construction of engines for high performance should be robust yet light in weight, and incorporate an adequate lubrication system both for the internals and bearings. Single-acting engines are universal for racing work, but the traditional double-acting engine is frequently used for less exacting duty.

To achieve lightness a fabricated construction is often chosen and there have been some very light engines produced by various constructors. Welding, or more frequently brazing, is often used to build up engine frames, steam chests, etc. One method much used in the past for piston valve chests, was to insert a cast-iron liner in a brazed-up construction of mild steel. This has the merit of allowing easy replacement of the liner when wear occurs. The interference fit should be small—say about '$\frac{1}{2}$ thou'—and combined with a good surface finish to avoid either difficult insertion and removal of the liner, yet avoid steam leakage.

The practice of drilling holes all over an engine to try and reduce weight is severely frowned upon! It is far better to modify the sections by milling flutes, or to choose a different material such as duralumin, where possible. Many successful engines have been constructed on petrol engine lines, having an enclosed crankcase; and it is quite probable that the more rigid arrangement improves mechanical efficiency enough to offset the increased weight.

Lubrication

Good lubrication is necessary for any engine, but absolutely vital for high performance jobs, otherwise serious failure is certain to occur.

Internal lubrication may be by small pump, geared in the region of 50:1 engine speed. Features of oil pumps will be discussed in Chapter 7. A good alternative is a type of 'hydrostatic' lubricator pressurised from the water feed line (*Fig. 7.11*). The familiar displacement lubricator is not really suitable for anything but moderate pressure flash plants, where the steam temperature is unlikely to be high enough to stop the condensing action. Where ball or roller races are employed for main bearings, big-ends, etc., splash lubrication may be sufficient, but plain bearings and big-ends are usually pressure-fed by a pump.

Superheat steam-cylinder oil has been the traditional lubricant for valves and internals, but several exponents have tried motor oil and even gear oil with success. 'Moly-slip' as an oil additive has been used with every sign of improvement in piston valve engines, and there are other additives available which may be worth trying. Heavy steam-cylinder oils have a high viscosity at normal temperatures, which may prevent oil pumps delivering when starting up. The time-honoured remedy for this is to warm up the oil container with a naked flame, but a better method is to by-pass some steam through a small heating coil before the engine is started.

It should be emphasised that modern steam oils are not necessarily of the traditional treacle-like consistency. We have experimented with several 'Mobil' steam oils which flow quite freely at room temperature yet are suitable for steam temperatures over 260 degrees C.

Engine Timing

A famous locomotive chief mechanical engineer is reputed to have said: 'any fool can arrange for admission of steam to a cylinder, but it takes a wise man to get rid of it'. From this it might appear that we are up against difficulties when choosing a valve arrangement! In fact, when high rotational speeds are concerned—and by this is meant 10,000 *rpm* or more—it is jolly difficult to get enough steam *in* to maintain a reasonable mean effective pressure, let alone the problems concerning the exhaust steam.

A brief explanation of the reasons for requiring certain features in valve timing is set out below.

(i) *Early cut off* Any two related quantities can be shown on paper by means of a simple graph. If the two quantities are constant a rectangular figure results (*Fig. 6.1*). Top left shows the rectangle

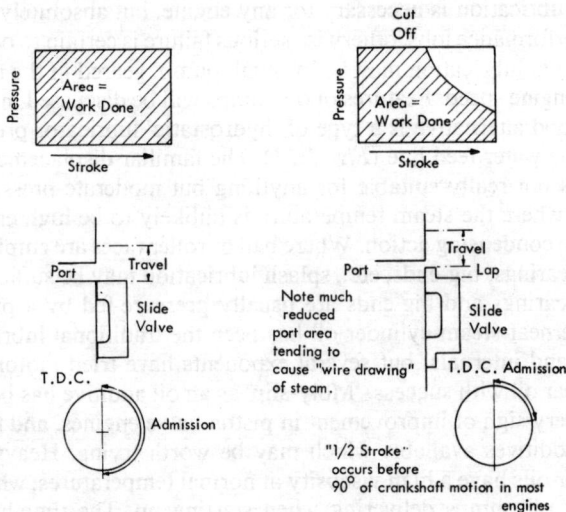

Cut
Off

Pressure

Area =
Work Done

Stroke

Pressure

Area =
Work Done

Stroke

Port | Travel

Slide
Valve

T.D.C.

Admission

Port | Travel | Lap

Note much
reduced
port area
tending to
cause "wire drawing"
of steam.

Slide
Valve

T.D.C. Admission

"1/2 Stroke"
occurs before
90° of crankshaft motion in most
engines

Fig. 6.1   Effect of early cut-off-valves fully open

produced by constant steam pressure if admitted throughout the stroke of an engine, and the area shown is proportional to the work done in one stroke. The right hand diagram shows the steam cut off at half stroke, the expanding steam results in a decrease of pressure over the remainder of the stroke, yet the total area is not much less than when steam was admitted for the full stroke. This results in steam economy and greater efficiency. Slide or piston valves worked by eccentrics or cranks must be longer than the span of the ports to achieve this, and this is known as *lap*. Valves and ports are shown below these diagrams, which are really theoretical *indicator diagrams* and found to be modified in shape if taken from actual engines. (Not possible as yet, from high-speed model engines).

(ii) *Lead (or advance)* It will be noticed that the diagrams show full pressure at T.D.C. although this is almost impossible to achieve by opening the valve at this point. Thus the steam admission has to be made a little earlier so that the steam has time to build up to full pressure. This is 'lead'— the amount that the admission port is open at T.D.C. If related to crankshaft angle it can be quoted in degrees of advance, just as in relation to the ignition point of petrol engines.

Readers are probably familiar with the type of timing diagram often shown in reference to the timing of petrol engines. *Fig. 6.2* shows typical diagrams for a single-acting engine using an eccentric driven

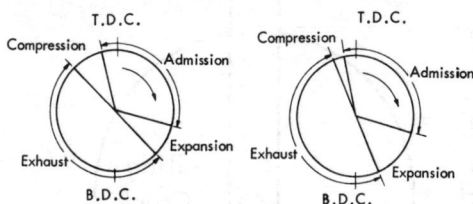

Fig. 6.2   Timing: (left) single valve and (right) two valves

slide or piston valve. The first thing to note is that the mid-opening point of the exhaust is exactly 180 degrees later than that of the steam admission. So if expansion of the steam is arranged, a similar compression period occurs after the exhaust closes. The diagram for separate admission and exhaust valves shows the improved timing obtained. What these diagrams do not show is the extremely slow opening and closing of the ports due to the type of motion given by conventional cranks or eccentrics *Fig. 6.3* attempts to explain this by showing a slide crank arrangement (sometimes called a *Scotch crank*) considerably enlarged; it will be noticed that the vertical movement is small approaching the dead centres—even for 90 degrees of rotation, and the movement is governed to some extent by the 'throw'.

The suggested travel on the diagram is ¼ of an inch: this gives a maximum opening of only ·037 of an inch. If smaller admission period is required—say about a total of 70 degrees, the opening is reduced to ·023 of an inch. This can be improved by increasing the valve travel, but even with ⅜-inch travel the maximum opening is only ·034 of an inch.

It is for the above reasons that locomotives, both full-size and model, use special valve gears to give quick opening and closing combined with adequate port opening. These valves gears operate by a linkage involving motion from the piston rod and eccentrics or followers, by which the cut-off can be varied to suit the conditions. The complexity of the arrangements renders them unsuitable for very high speeds—for model locomotives *rpm* are usually reckoned in hundreds only, whereas for most racing hydroplanes 8,000 *rpm* are on the low side.

In many high-speed engines attempts are made to avoid the use of simple cranks or eccentrics in favour of cams or devices which speed up the motion of the valve at the required point. A variety of valve

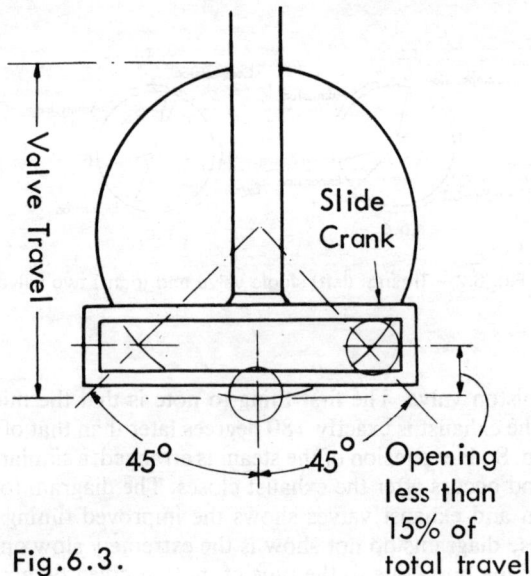

Fig.6.3.

Valve Travel

Slide Crank

45°   45°   Opening less than 15% of total travel

Fig. 6.3   Valve opening for 90 degrees admission

gears for both single cylinder and other engines, all of which have been used in model flash engines at some time or other, are shown with the following notes. A complete discussion of the pros and cons of each arrangement is not possible here, but a few notes on each of those illustrated may be of interest.

Firstly, the valves themselves—if the valves involve sliding surfaces—they all are likely to suffer (even when made of special materials) if the temperature much exceeds 400 degrees C. Those which are not pressure balanced, e.g., a slide valve, are likely to suffer from excessive friction.

High temperature is difficult to control when a flash boiler is working flat out, and at least two examples are on record of twin-cylinder, piston valve engines that have employed water cooling of the valve chest—a brief description of one of these will be given in Chapter 12. A form of heat exchanger or feed-water by-pass would appear to offer greater efficiency than the water feed direct from scoops, as used in both of the examples mentioned.

The poppet valve or similar kind is most attractive because of its self-sealing capabilities, and the fact that the valve head requires no lubrication in order to make a very effective seal. Two valves are required, however, unless the exhaust is of the 'uniflow' type. Since the valves operate at engine speed, valve bounce may occur even when loaded by springs in addition to steam pressure.

The valve gears are next considered.

Type A   Vertical valve driven direct from mainshaft by eccentric or

Plate 14   A typical single-cylinder engine with vertical piston valve (J. Benson)

small crank: this is good mechanically, eccentrics can be balanced by weights in a similar manner to piston and crank arrangements. Steam passages fairly long. See *Plate 14*.

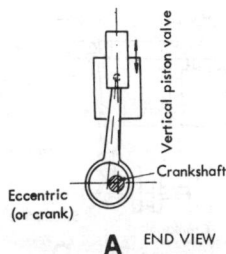

A   END VIEW

Type B   Bevel gears driving vertical shaft, valve horizontal. Good mechanically with suitable bevels, well lubricated. Twin cylinder engines sometimes known as the 'Westinghouse' type from high speed generator engines used in the past. Short steam passages, 'Banjo'-type strap mostly used to avoid excessive angularity of the valve rod if operated more directly.

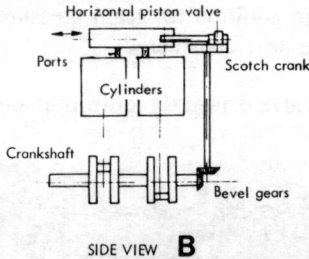

SIDE VIEW   **B**

Type C   Requires an articulated joint at the lower end of bell crank plus two other joints, mechanical losses may not be significantly greater than *B* if well made, but some lost motion is likely. The original Westinghouse engine had this gear.

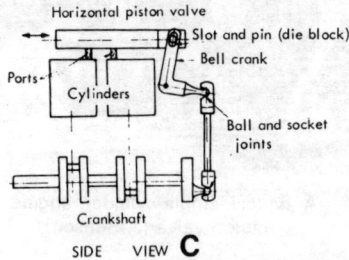

SIDE   VIEW   **C**

Type D   Swash plate drive, a simple drive giving harmonic motion. A problem lies here in the method of engaging swash plate to give good wear. Suggested method by E. T. Westbury (*Fig. 6.4*).

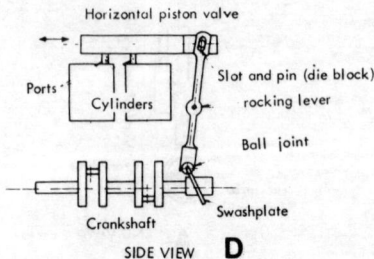

SIDE VIEW   **D**

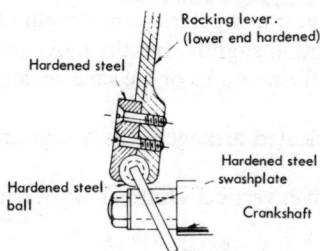

Fig. 6.4 Swashplate (E.T.W. design)

Type E Shorter steam passage than vertical valve but greater inertia due to more parts. Rocking couple if centre line of valve is out of line with the drive. Suitable for single cylinders only.

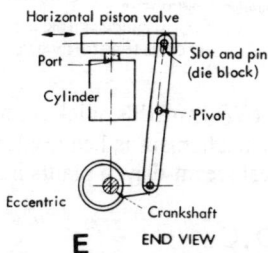

Type F Semi-rotary. Long ports can be used to offset the sluggish motion at the extremes of movement. Can be pressure-balanced if required. Good mechanical drive but not too successful when tried in the past.

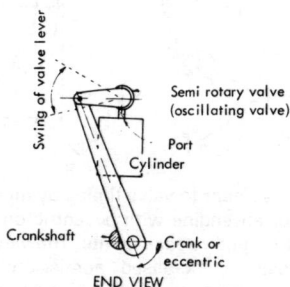

In all these valve gears, angularity and length of links or eccentric straps modify the motion slightly, but the movement given is approximately harmonic. Off-set cranks or 'de saxe' arrangement can modify timing slightly.

The more sophisticated arrangements are given below.

Type G   Piston Valve, vertical (Jutton), or overhead (Pilliner). The

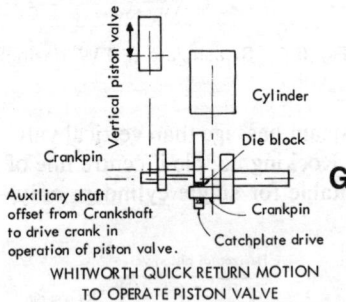

WHITWORTH QUICK RETURN MOTION
TO OPERATE PISTON VALVE

valve gear is a form of Whitworth's quick return motion and speeds up the inlet timing. The mechanism is heavily loaded, but operates well, since two of the fastest steam-driven plants have had this gear applied

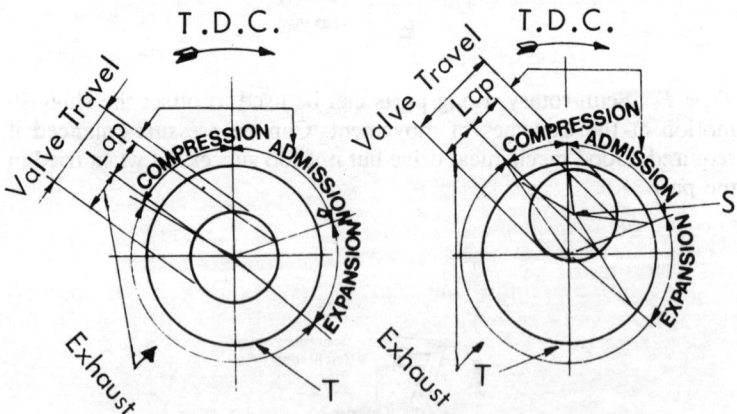

Fig. 6.5   Improvement to valve timing by means of a 'quick return' gear, for an engine with eccentric on the camshaft admission, cut-off and release points. The valve diagram for the *Frolic* indicates increased admission opening and reduced compression period. Arrow marked S indicates centre of eccentric shaft relative to crank centre, arrows marked T indicate the crank circle

to single-cylinder engines. A small die block on the return crank is essential, otherwise only line contact occurs and wear is then excessive. Valve timing is improved depending on amount and angularity of the offset. The timing diagrams of Bernard Pilliner's single-cylinder engine before and after fitting this gear are shown in *Fig. 6.5*.

Type H Poppet valve—cam operated, valve mechanically robust, self-sealing without lubrication. Normally any timing possible for one purpose—admission or exhaust—not both. Two valves are employed if exhaust is to be controlled as well as admission. The cam contours needed give rise to problems in valve bounce, since the valve operates at engine speed and the acceleration is fierce if short periods are required. Loading with springs, in addition to steam pressure, seems to improve matters. A pre-war exponent, Leslie S. Broad, was well-known for his three-cylinder poppet valve engines which put up some fine performances for the time. A twin-cylinder engine by the same exponent was reported to have attained 15,000 *rpm* on a bench test, but this was a small engine and the valves would be very light.

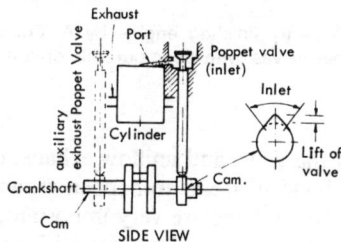

H

The force required to lift a poppet valve against high steam pressure can be considerable and the valve gear can be highly stressed—one engine was known to lift the cylinder head rather than open the valve!

The 'Spartan' poppet valve engine, designed by E. T. Westbury, is a straightforward and easily constructed engine very suitable for flash steam. Drawings are available for this engine from Model and Allied Publications Ltd. A new engine by A. Cockman with 'desmodromic' valve operation is shown in *Plate 15*.

Type I Disc valve operated by piston. This is a variation of the poppet valve without the use of a cam and associated mechanism. A short admission period is usual and this tends to build up high boiler pressure. Engine difficult to start if only 'uniflow' exhaust system is used—a separate exhaust valve makes things easier. Admission of steam occurs equally before T.D.C. as after, and a timing diagram of

Plate 15   A partly finished engine by A. Cockman. Two
poppet valves with desmodromic operation

the events given by the valve and uniflow exhaust does not look very
convincing. 'Valve bounce' seems to happen around 8,000 *rpm*. The
materials for the valve and peg are very important, and Jim Bamford
had to use motor-cycle exhaust valve material for the disc, and nickel
steel for the peg. (*Fig. 6.6*).

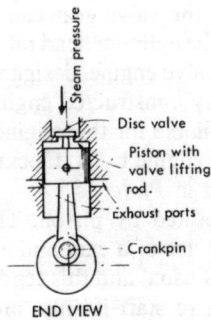

END VIEW

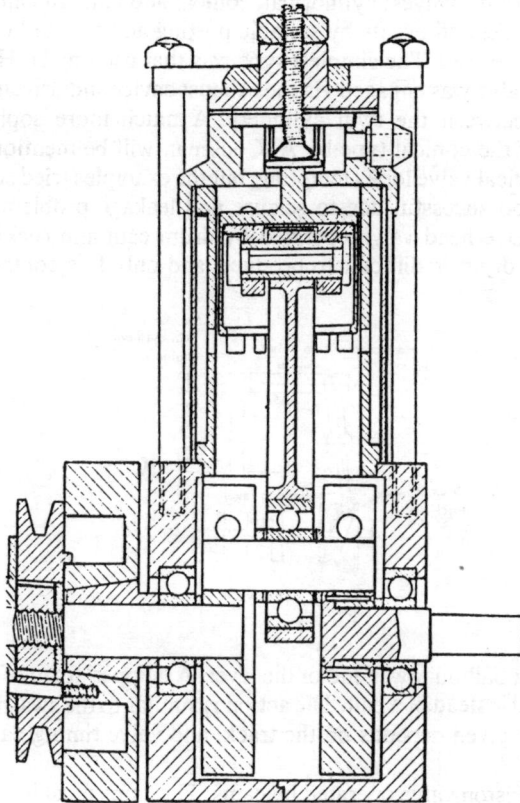

Fig. 6.6   Piston-operated disc valve engine (J. Bamford),
showing the general arrangement of the engine, in section

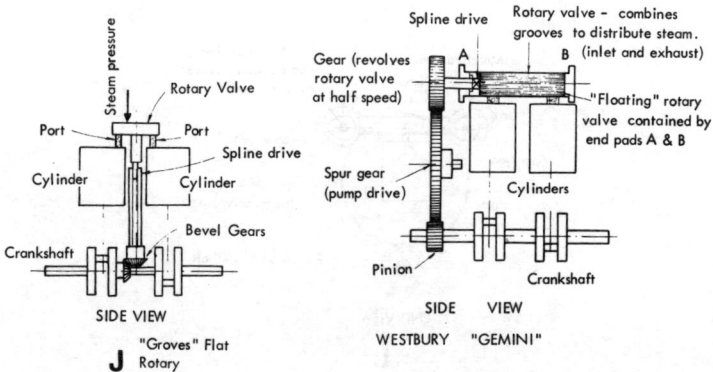

Type J   Rotary valves; cylindrical, conical and flat. All can be made to give any desired events by suitable porting and all can be pressure balanced if desired. The simplest type was that used by H. H. Groves where the valve was engaged via the exhaust cavity and driven by bevel gears set between the twin cylinders. A much more sophisticated example of the conical type by A. Cockman will be mentioned later. The cylindrical valve looks tempting, but the examples tried so far have not been too sucessful due to seizure and leakage problems.

Type K   Overhead valve operated by drum cam and rocking lever. 'Track' on drum is difficult to produce, and only line contact results

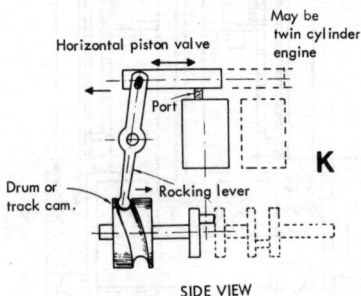

from use of ball on lower end of the lever. A 'slipper' form of die block can be used instead of a ball. The action is similar to the swashplate, but the motion given depends on the track, and valve timing can thus be improved.

Type L   Piston valve operated by cams. The cams must have complementary contours and driven in step. Very few examples are known of this type. Two engines were made by F. Lowne many years ago. A single-cylinder engine had two overhead valves operated by camshafts

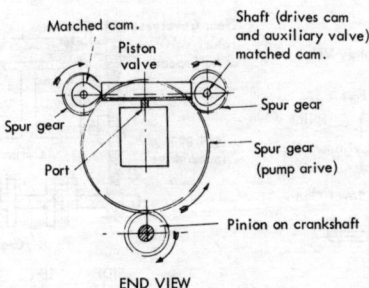

Plate 16 Two engines were designed and built by F. Lowne, having piston valves operated by complementary cams. This is the single-cylinder type

Plate 17 The three-cylinder radial engine built by F. Lowne

on either side. Although having only $\frac{9}{16}$-inch bore and stroke, a very good performance was obtained with a hydroplane of lightweight construction. A more ambitious engine was a three-cylinder radial type having larger cylinders, each with single overhead valve operated by bell-cranks and push rods from two cams only (*Plate 17*).

The uniflow principle

The term 'uniflow' really applies only to engines having exhaust ports in the cylinder wall opened by the descending piston, with no other method of exhaust. The path of the steam is in one direction only, entering by the admission valve, expanding as it operates the piston then passing through the uniflow ports towards the end of the stroke.

In full-size engines using drop valves, which provide very early cut-off and thus large expansion ratios, very high efficiency is obtained. Unfortunately, such valves are unsuitable for high speed, and in model work straight uniflow engines have in most instances been provided with poppet valve admission. The sudden opening of the ports clears the cylinder effectively in a similar manner to a two-stroke petrol engine, but the clearance volume left when the piston is at T.D.C. affects the running and power output. Compression must occur and it is this volume which controls the maximum compression; the calculation is the same as that for petrol engines:

$$\text{compression ratio} = \frac{\text{swept volume} + \text{clearance volume}}{\text{clearance volume}}$$

If an engine had swept volume = 1 cubic inch, clearance volume = 0·2 of a cubic inch, the calculation would give $(1 + 0·2) \div 0·2 = 6:1$. Now if Boyle's law is applied, the compression pressure will thus be about 6 times atmosphere, approximately 6×15 *psi* absolute. It is thus obvious that the engine will not operate on less than 90 *psi* absolute. This calculation must not be regarded as accurate, since it does not take into account temperature changes nor the fact that the gas being compressed is steam and not a perfect gas, but it does give an illustration of what occurs.

Jim Bamford found that altering the compression ratio on one of his uniflow engines produced a remarkable increase in power during brake tests. The change was from 5 : 1 to 5·7 : 1, and the power produced was 1·2 B.H.P. at 8,000 *rpm*, about 100 per cent improvement! This engine was fitted with the piston-operated disc valve, which, of course, does give very early cut-off and it is interesting to note that the steam consumption of 20 lbs. per brake horse power hour compares well with full-size engines of simple, non-condensing type.

From all this it may be inferred that model uniflow engines are seldom found in plants for moderate performance, for they prefer high pressure and fairly high revs. Incidentally, the surmise that steam plants are nice and quiet when compared with small petrol engines is a complete fallacy! A small high-pressure uniflow engine is every bit as noisy as its two-stroke counterpart. M.P.B.A. rules require the silencing of all racing steam hydroplanes, irrespective of the type of engine used.

At one time the provision of auxiliary ports in the engines of racing boats was universal practice among the flash-steam fraternity, even when other means of exhaust were used as well. Some exponents have raised doubts as to the effectiveness of these ports on the grounds that: (a) blocking off such ports makes no noticeable difference in performance. (b) the ports tend to shear the oil film (if any!) and (c) the

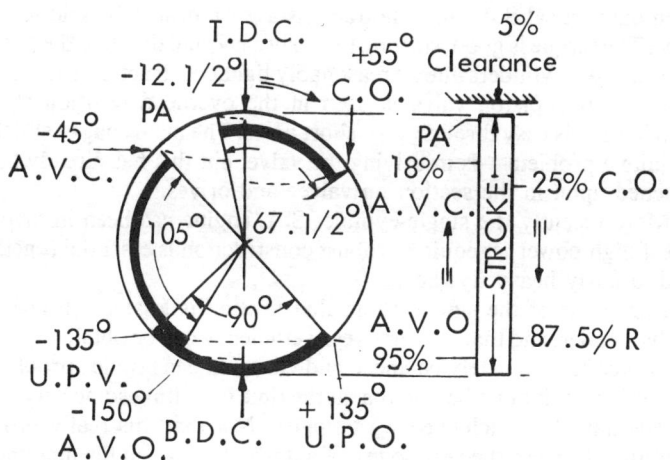

Fig. 6.7 'Uniflow' timing with auxiliary exhaust valve
(Thomas Hindle)

collection of the exhaust steam around the cylinder walls presents difficulties and adds unnecessary weight. These points appear formidable, but it should be remembered that the engines in question had valve gears which may not have given timing best suited to proper uniflow operation, and that the alternative exhaust system was fairly free.

D

A well-known pioneer of flash steam, Mr. Thomas Hindle, advocates a small auxiliary relief valve to avoid excessive compression yet a very small clearance volume. The auxiliary valve should not open until the uniflow ports are about to close so that the main exhaust really is 'uniflow' (*Fig. 6.7*).

It is interesting to note that one of A. Martin's *Tornado* series successfully used a poppet-valve uniflow engine equipped with a small relief valve of the piston type of $\frac{5}{32}$-inch diameter, but with the exhaust valve opening earlier than advocated by Mr. Hindle. This craft was a lightweight job weighing only $4\frac{1}{8}$ lbs. and often lapped the course around the 40 *mph* mark—a top performance in the days of fully submerged props.

Multi-cylinder engines

From the earliest days, the traditional engine for a racing plant has been the vertical S.A. twin and there is a great deal to be said for the type. The torque is good, starting up fairly easy, and dividing the power between two cylinders allows reasonably light construction and a light flywheel. If a piston valve is used in the overhead position steam distribution is easy through very short ports. The main snag—and this is quite a problem—is in driving the valve, but this has already been touched upon in the section on valves and drives.

More recently the single-cylinder S.A. engine has been in favour, but if high power is required, robust construction is essential together with a fairly heavy flywheel.

Exponents of the single, claim that friction is less, steam leakage paths smaller, and that greater overall efficiency can be obtained; but it can never be completely balanced and vibration can provide problems. Nevertheless, it must be said that at the time of writing some very high speeds have been achieved with singles. Not that this really proves anything, because the real power of a flash plant lies in the amount of steam evaporated by the boiler and lamps, and providing the engine uses the steam efficiently the number of cylinders are perhaps not too important, but there is little doubt that multi-cylinder engines are attractive. If correctly designed better balance and perhaps full power at higher *rpm* may be obtained—a desirable feature for modern racing hydroplanes.

It should be noted that the vertical twin with cranks at 180 degrees is *not* in perfect balance, although to the uninitiated it may appear so. Static balance is obtained, but the pistons operating in different planes cause a 'couple', the magnitude of which partly depends on the distance between the crank centres. An improvement is made by attaching balance weights to the outer webs only. Motorists who have watched their road wheels being balanced on a wheel balancing machine will have some idea of the subject, and know what happens when the wheel is rotated after initial balance for static forces only.

Vee-twin engines with cylinders at 90 degrees are much better, since the primary forces can be balanced out by counter-weighting the crank. The method is to balance all rotating weight and reciprocating weight of *one* piston and rod. A vee-twin engine by Jim Bamford is shown in *Plate 18*.

Plate 18 Vee-twin engine with piston-operated disc valves, by J. Bamford

Engines employing more than two cylinders are few and far between. In pre-war days the 3-cylinder poppet valve uniflow engines constructed by Leslie S. Broad (Cardiff) were very successful. Several similar engines were built by his fellow club-members. Apart from the improved torque, balancing is still a problem, and such engines have found little favour apart from one or two examples in cruising boats.

In the case of four-cylinder engines only two racing examples are known. The engine constructed by J. Willis (Dublin) was a flat four with piston valves operated by track cams. It was this plant that on test recorded only 25–30 *psi*, and convinced some enthusiasts that high pressures were not obtained in model flash plants! In truth, as men-tioned elsewhere, the engine steam consumption controls the working pressure, and it is apparent that this particular example resulted in the four cylinders using the steam almost as fast as it was generated.

The vee-4 layout has interesting possibilities, since if arranged with cylinders at 90 degrees and crankshaft at 180 degrees, primary balance is excellent if the mass of one piston and con-rod is balanced by bob weights on the crank.

A pioneer of flash steam, George Noble (Bristol), constructed a very nice vee-4 engine in the early 'twenties. Although of the slide-valve pattern the design was such that conversion to piston valves would present little trouble. As recently as 1969. Edgar T. Westbury drew attention to the possibilities of the vee-four. He outlined the design mentioned in Chapter 2, which could be made in a size suitable for model hydroplanes if required. The 'floating' rotary valve solves problems of side thrust that have caused excessive wear on some cylindrical rotary valve engines.

In 1928 Edward Hobbs, associate of the Institute of Naval Architects, published a design for a six-cylinder opposed engine in *Wonderful Models*. This also has a rotary valve, but as far as is known no examples were ever produced. The design evoked some criticism at the time on the grounds of lubrication, use of slide cranks and other features.

Although perhaps a little beyond the scope of this chapter, some mention should be made of a different type of prime mover—the steam turbine. A few articles have appeared at various times in *Model Engineer* concerning turbines, but not many have been flash steamers. Two notable exceptions were those by Prof. D. H. Chaddock and J. A. Bamford. The latter constructor's efforts are described in Chapter 12, as the experiments formed part of the development of a series of flash-steam hydroplanes for competition work.

Full-size turbines reach very high efficiency and this is achieved by making the steam pass a series of turbine blades before exhaust. Some early turbines are quoted as having no less than 30 stages before condensing. The point is that the high pressure and highly superheated steam is expanded progressively right down to sub-atmospheric pressure in the condenser, thus exchanging a lot of energy. As the size of the turbine is reduced so the efficiency tends to fall. When model sizes are reached this seems well below that of the better racing engines working on early cut-off.

The model turbines constructed to date are of the single-wheel pattern. As far as can be ascertained no 'staged' model turbines have yet been attempted.

Plates 19 and 20 illustrate a well-known example by Prof. D. H. Chaddock and were intended to provide information leading up to model gas turbines. During the course of this project a replica of the

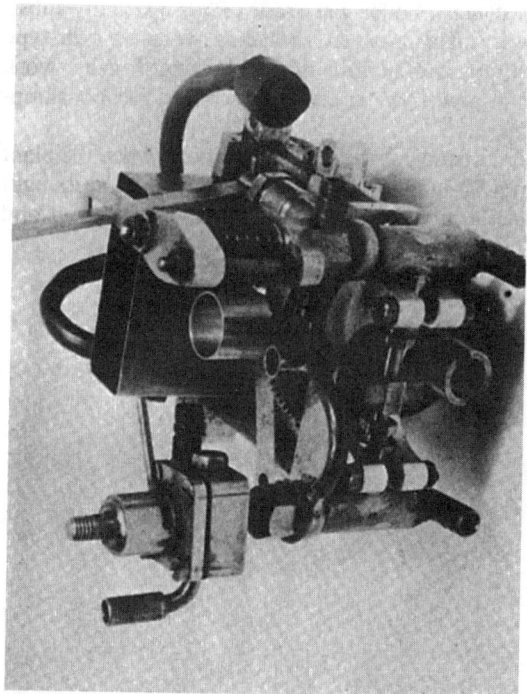

Plates 19 and 20   Professor D. H. Chaddock's experimental steam turbine

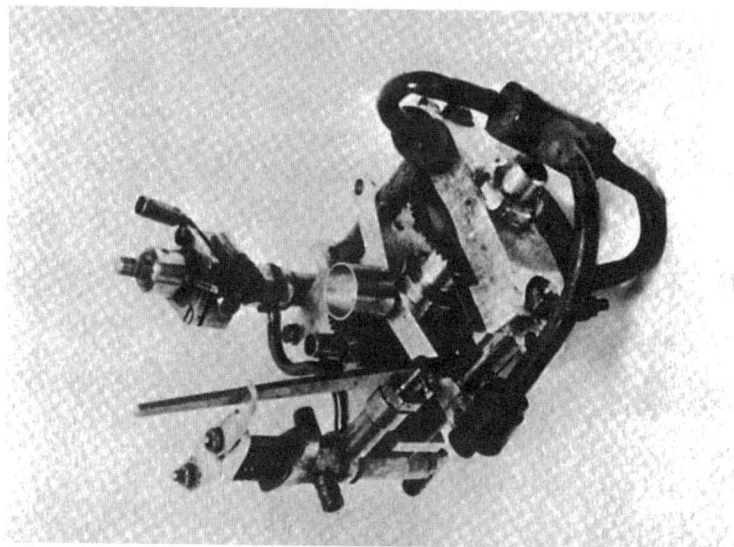

'Groves' model aircraft flash boiler was made (*Plate 8*). Many subsidiary experiments were also involved, including work on gear-type pumps, oscillating pumps and methods of gear-cutting. Further work included bench tests of power, water consumption and fuel consumption.

We well remember assisting at the Blackheath pond when the plant had been installed in a 24-inch hydroplane and marvelling at its light weight of 3 lbs. and the meticulous workmanship of the plant and installation.

CHAPTER SEVEN

PUMPS

Engine power is used through reduction gearing to work feed water, and often oil and fuel pumps, to operate the flash-steam plant. The correct amount of each fluid supplied is most important. Possible means for metering will be considered under each system.

The boiler feed pump

The duty of any boiler feed pump is merely to balance evaporation. In conventional boiler feeding, the output of the water pump is usually arranged to provide a surplus of water. The 'balance' is controlled by regulation of a valve within the delivery pipe, which spills some of the water, and returns it to the feed water tank, or in the case of most marine installations, overboard. Such a system works reasonably well if the boiler pressure does not fluctuate and if the loss of steam through safety valve and auxiliaries is moderate. The operator of such boiler plant, however, has a reserve of water available by use of the emergency hand pump, he is also aware of boiler contents by observation of the water gauge.

Operation of a flash boiler does not always permit the opening and closing of a by-pass valve to spill water. Spillage of water through this valve will only remain constant for a regular boiler, or to be more correct, steam generator pressure. A rise in pressure provides more resistance at the check valve resulting in increased spillage; the water taking the path of least resistance. It may equally be understood that a drop in pressure permits more water to enter the boiler tubing.

Insufficient water passed to the coil of boiler tube will not generate the quantity of steam required, but may superheat the steam beyond the temperature at which the engine can safely operate. Delivering water to the boiler in excess of heat applied promotes low steam pressure without superheating; wet steam is generally of little use. In fact, flash plants having no specific water or temperature controls tend to be unstable, either the steam is too wet or too hot!

Mention should be made here of the pressure relief valve used by Prof. D. H. Chaddock on one of his experimental steam turbine plants. It consisted of a spring-loaded ball valve in the pump feed line loaded to blow at 500 *psi*. Using turbine nozzles of calculated proportion the arrangement provided a considerable measure of control. It was found necessary to fit a check valve between the relief valve and the boiler. Without it all the water in the tubing tended to be lost when the valve operated, rather than merely surplus feed water being eliminated.

J. Bamford has used a similar valve on his latest flash steamer, reporting that it prevented the typical slowing towards the end of the run. This is interesting because in theory the arrangement should not work with a reciprocating engine plant.

(In a turbine plant the steam consumption is governed by the nozzles and is quite independent of the rotor speed and thus the pump speed, which is not the same as when a reciprocating engine is used.)

Plate 21   H. W. Saunder's water feed control (left). The engine is built to E. T. Westbury's 'Trojan' design, with pumps and gearing added

Another interesting method of control is used by H. W. Saunders (St. Albans) in a moderate performance plant (*Plate 21*). In this instance, the inlet to the feed pump is deliberately restricted by means of a ball valve, the lift of which is limited by a simple screw control.

This device works well; once set, speed can be adjusted by the heat supplied, and the plant will continue running as long as the fuel lasts. As far as is known this arrangement has not been adopted by hydro-plane exponents, but it might solve many problems if it would operate at the high pump speeds required.

An ideal water pump should be capable of exact metering. Indeed, some past flash-steam plants applied to model work incorporated levers and linkage to vary the travel of the ram, even while working. Simplicity and reduction of weight has always to be considered in mechanical systems, alternative means can be applied to vary output from the pump. As the bore of the pump is constant, variation of the movement to the ram is the only method available to vary output. To vary the stroke of the ram, a connecting rod with a screwed-in crankpin is applied to tappings of differing radii within a crank disc (*Fig. 7.1*).

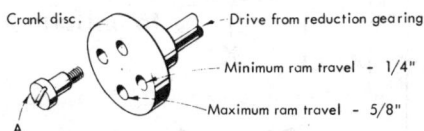

Fig. 7.1   A pump-driving crank illustrated at about half-size. The crankpin, indicated by A, is screwed in at various radii

It is most important that the ram travels the length of the barrel, any dead space between the end of the ram and the delivery valve could trap air, forming a cushion and reducing effective working. It follows that variations of the stroke by alteration of the driving crankpin, must also provide for bodily moving the pump and flexing of pipework to avoid

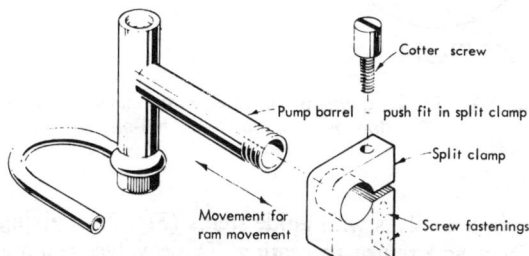

Fig. 7.2   The pump support clamp is shown at approxima-tely half-size. The curved pipework is left soft to permit flexibility of banjo union swivels

dead space within the pump barrel. The mounting to provide for alteration of pump position may be provided by a simple split clamp, as shown in *Fig. 7.2.*

Other important design features should provide for accurate fitting of the ram, either with ram packing, 'O' rings or a screwed packed gland, to resist leakage and operate with minimum friction. Perfect sealing of both suction and delivery valves is without doubt of utmost importance. The valves used in past flash-steam pumps have taken several forms. The wing type (*Fig. 7.3*), which, with some variations

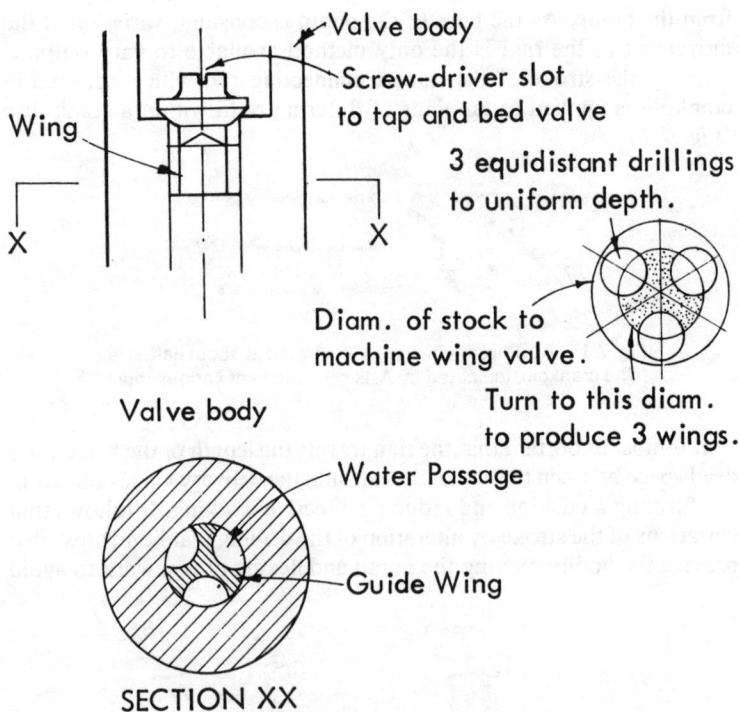

Fig. 7.3   The wing type pump valve with details for drilling.
The drawing is about twice full-size

turns into poppet valves with guide stems (*Fig. 7.4*); giving accurate seating, lifting and return to seating. These valves require machine operations and turning of small diameter stock to accurate limits if they are to be trouble free. It is thus that the ball-type valve is more generally employed within model and small pumping mechanisms. A large selection of both stainless steel and phospher-bronze balls are readily obtainable.

Pumps in racing flash-steam plants work rather fast even though geared down. Some means of assisting rapid and complete closure of the valves should be incorporated in the design of the pump.

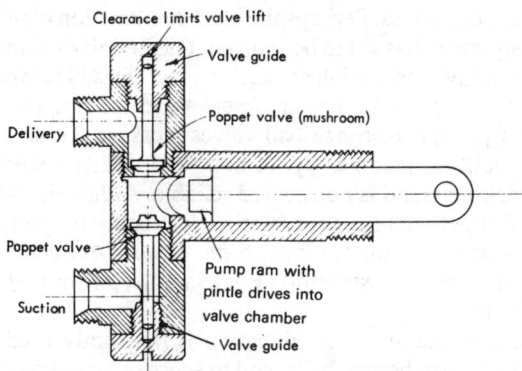

Fig. 7.4   The feed pump with guided mushroom (poppet) valves, approximately full size

With ball valves, the lift or rise of each valve must be controlled. Some means should be provided to prevent side movement, without restriction of the waterway. This may be achieved by fitting a cap over each valve or providing a cage around the ball, as in *Fig. 7.5*.

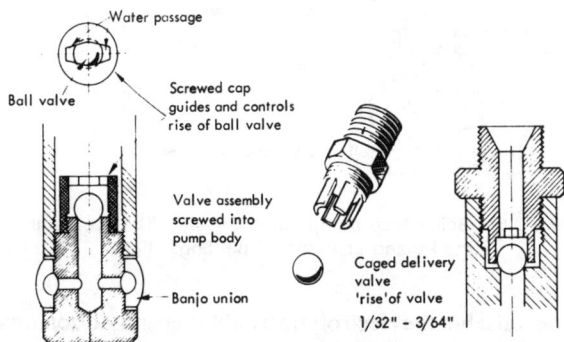

Fig. 7.5   Guided ball valves

The selection of materials used in the construction of the water pump is important. Some dissimilar metals or alloys united through water promote electrolytic action, often resulting in pitting or wasting of the base material. Several constructors have used, and still continue to use, aluminium alloy for water pumps in order to reduce weight. More than a few headaches have resulted when the pump failed due to oxide fouling the valves. Perhaps the ideal combination of materials for pump construction has yet to be evolved. Plastic valves, non-corroding aluminium alloy and stainless steel appear feasible, working into stainless steel pipework. For a *simple* water pump, we advise the adoption of phospher-bronze ball valves working within a gunmetal casting or built-up pump body. A hollowed stainless steel ram with screwed packing gland is considered reliable. Although stainless steel balls would appear to be the automatic selection for the valves working upon brass or gunmetal seatings, some electrolytic action has been experienced by past constructors. For example, pits or craters form in the steel balls.

Despite this, the above combination is frequently used for racing plants, as phospher-bronze balls tend to score or go out of shape under extreme conditions. The seats and ball valves may be inspected at

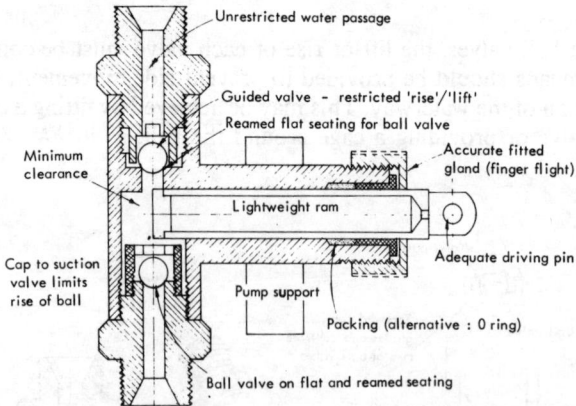

Fig. 7.6   Boiler feed pump about full-size. The accurately fitted gland at A should be 'finger tight'

regular intervals, but the electrolytic trouble depends to some extent on the purity of the water available.

The best method of seating ball valves is, first, to solder a ball of the same diameter on a piece of tube of slightly smaller size. This is rotated in a drill chuck and the valve seat burnished. No abrasive is required and less than a minute of this treatment will do the trick.

The most perfect feed pump can fail if microscopic debris passes through with the water. A suitable filter must be incorporated, either in the pick-up system for marine installations or within the feed water-tank: details of filters are mentioned elsewhere.

The basic design requirements of a water pump are indicated in *Fig. 7.6.* For a pump with a working ram speed of say 4,000 strokes per minute, operating direct from the engine crankshaft the design indicated in *Fig. 7.7* is worthy of consideration. The design has been

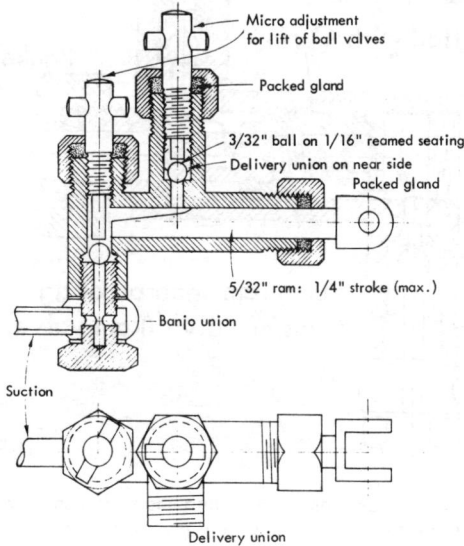

Fig. 7.7   The high-speed boiler pump, shown approximately full-size

applied to a well-known steam plant driving a power boat. Accurate control of each ball valve is provided by the stems working through packed glands. Provision of a uniflow pump in the larger size may be made by incorporating a transfer valve with the ram (*Fig. 7.8*). Water is drawn through the traditional suction valve and passes through the ram valve as the operating gear drives the ram forward. The reverse travel discharges the contents of the barrel through the delivery valve while drawing in another charge of water. This form of pump relies on a well-fitted ram, this may now be done with the well-known 'O' ring.

Past constructors had to rely on graphited packing or leather cup-washers. The fitting of the gland to this pump, and indeed, all pumps, must be accurately carried out; the inclusion of a pressure sleeve within the gland is advised. The assembly should only be tightened

Approx. size indicated 7/16" bore 9/16" stroke

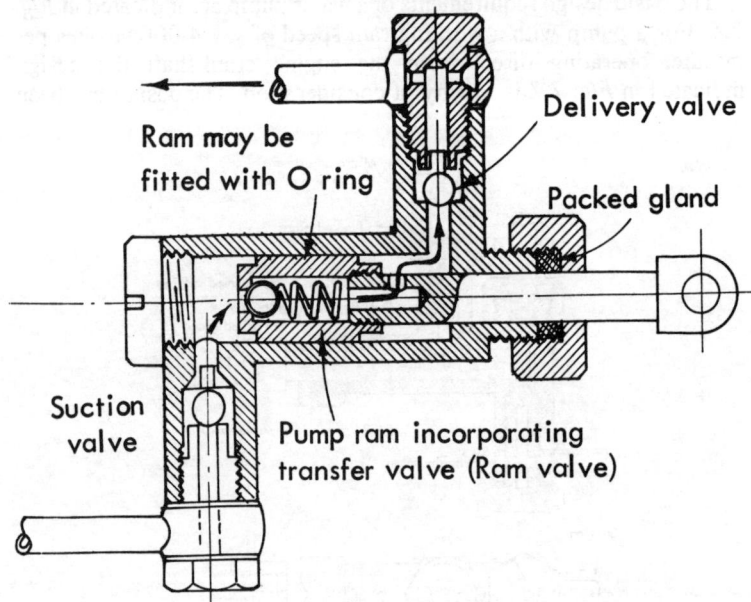

Fig. 7.8   Uniflow boiler feed pump shown full-size. Water enters the pump through the pipe at the bottom

sufficiently to compress the packing and prevent leakage, a knurled edge to the gland nut is often sufficient for finger operation.

To start up a flash-steam plant it is first necessary to pass pass some water to the boiler to generate steam and turn the engine. An auxiliary hand-operated pump forms part of the steam plant. To reduce weight and complications, some racing hydroplanes incorporated a combination pump, two barrels and rams working into a common chamber of suction and delivery valves. Naturally, only one ram can operate at one time. After the initial generation of steam and revolving of the engine, the barrel and ram of the auxiliary pump has to be isolated from the valves. *Fig. 7.9* indicates a typical arrangement.

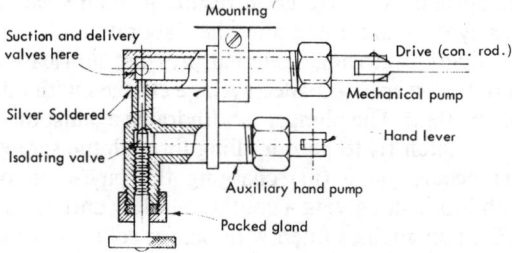

Fig. 7.9 Feed pump with starting up auxiliary pump, shown half-size

Oil pumps

Lubrication of steam engines applied to power boats often relies on the simple gravitation of oil from a small header tank to oil cups on bearings and other parts. This is quite satisfactory for many bearings. A gravity system of big-end lubrication applied to high-speed, double-acting engines has also worked well. Some applications, however, require more positive means of oil distribution.

Motor-oils—even multi-grade types—tend to run through the smallest pipes very quickly when working temperature is reached. Delicate restriction is a necessity. For this reason, model steam engines often have higher viscosity oils for lubrication of bearings. These oils also present problems, for no oil may flow until the engine warms up! Oil feed from a header tank is unlikely to be steady due to these variations in

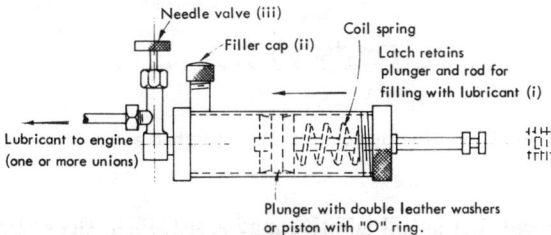

Fig. 7.10 Simple pressure lubrication for the bearings and external parts

viscosity relating to temperature. Some bearings are starved, while others tend to receive more than their fair share—friction in small-bore pipes plays funny tricks.

A simple system that gave good results in both steam and petrol engines in early days did not require the elaboration of oil pumps and associated reduction gearing. External pressure lubrication required nothing more than a spring-loaded syringe charged with oil. *Fig. 7.10* indicates the method. The plunger is withdrawn against the spring and retained by the latch (i) to permit filling through the screwed cap (ii) opening the needle valve (iii) charging the pipes, oil outlets and bearings with lubricant, giving a continuous flow until the syringe was exhausted. Certain engines fitted with ball or roller bearings may not require pressure lubrication for externals if only working for short periods, as in racing. For example, the late Bernard Pilliner arranged crankcase oil feed by the centrifugal force set up by the circular course. Application of lubricant before each period of running the engine is

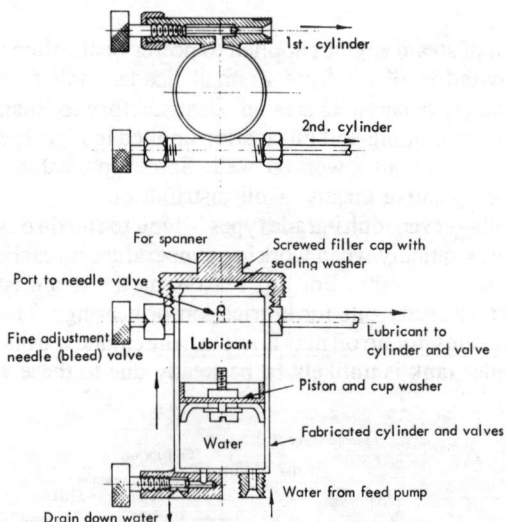

Fig. 7.11   The hydrostatic lubricator, half actual size

often sufficient, but an oil additive may assist. Plain sleeve bearings, eccentrics, big-end bearings and other portions of the engine are best served by oil delivered under pressure, through drillings in the crank-shaft and main bearings. Positive feed may be provided by various forms of force pumps. The pump provides lubricant on the total loss system through ports in the crankshaft, where a small annulus turned in the shaft maintains oil feed at all degrees of rotation.

Internal lubrication of piston and valves may be provided by a

similar force pump working through a check valve, very much like the
boiler feed pump. The displacement lubricator is not suitable for an
engine working on superheated steam, as condensation of steam
cannot take place. Some past constructors have endeavoured to overcome this problem by cooling the lubricator with fins and providing
large radiation surfaces; little success has resulted from such experiments. The term hydrostatic can truly be applied to the internal
lubrication system favoured by some exponents of the flash-steam
hydroplane; it takes the form of a hydraulic system using some of the
feed water to force lubricant to the internal parts of the engine. It will
be noted from *Fig. 7.11* that the cylinder of lubricant is fitted with a

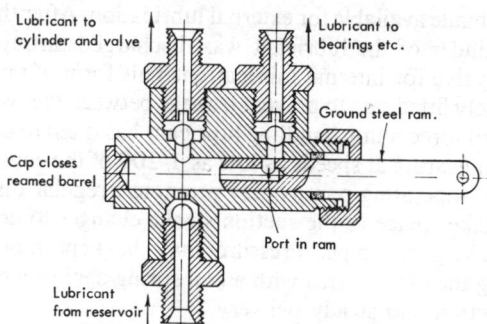

Fig. 7.12   Combination lubricant pump, actual size

piston and cup washer, although some exponents use a simple loose-
fitting piston only. Filling is by the removal of a screwed cap. Pressure
of the feed water being in excess of cylinder steam pressure displaces
the cup washer and lubricant dependent on the opening of the needle
valve. The system may well incorporate two or more control valves in
provision of lubricant to possible multi-cylinders, valves, etc. Delicate
metering of lubricant is possible, and may even be applied to external
use if a suitable grade of oil can be used. If two grades are really
necessary, two such hydrostatic devices may be brought into use to
avoid the necessity for additional pumps. To release the water from
below the cup washer, a drain valve may be provided, although
downward movement of the piston forces water back into the boiler
feed pipeline. Some objection may be raised to this on the grounds that
oil, leaking past the cup washer, may find its way back to the valves of
the water pump.

The oil pump or pumps may be smaller versions of the boiler feed
pumps; operated in a similar way but at slower speed; after all, the

quantity of oil required is rather small when related to water. It is often beneficial to instal separate pumps for internal and external lubrication, each pump taking oil from separate vented tanks charged with the required grade of lubricant. A very successful oil pump applied to a hydroplane delivered common lubricant in two stages through independent pipe-work by the movement of a single ram. The pump involved the addition of another delivery valve and minor alteration of the ram. By study of *Fig. 7.12*, it will be seen that oil is drawn into the barrel in the usual way. Forward movement of the hollow ram passes some lubricant to the port-controlled delivery valve, where pipework distributed the oil to bearings, etc. The size of the port, related to the total movement of the ram, determined the proportion of pump barrel contents made available for external lubrication. After the port closure, the remainder of the lubricant was discharged through the primary delivery valve for internal lubrication. This form of pump relied on a very closely fitted ram to prevent leakage between the two systems.

A small force pump handling lubricant has often to operate through reduction gearing at speeds as low as 1/200 of the engine revolutions. Such slow-operating pumps may promote irregular distribution. The pulsing takes place as the suction stroke changes to delivery-pushing the oil along the pipe. Pressure may be kept more constant by operating the ram, loaded with a coil spring against a cam, to provide rapid suction and steady delivery.

Fig. 7.13 indicates the cam with characteristic shape—often termed *snail cam.* pumps operated this way should be well supported, the ram

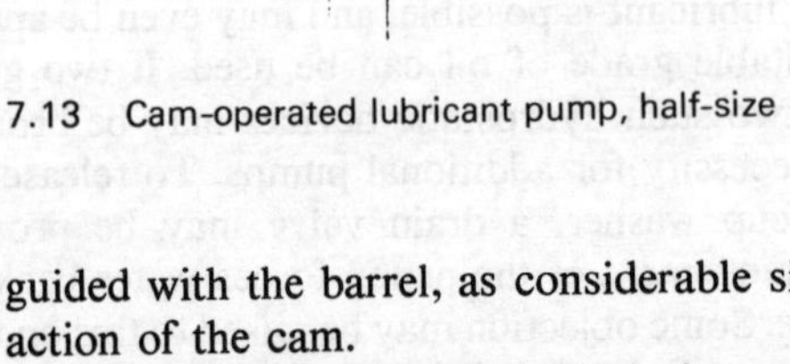

Fig. 7.13   Cam-operated lubricant pump, half-size

well fitted and guided with the barrel, as considerable side thrust is exerted by the action of the cam.

The small oscillating pump, submerged within its own supply of lubricant and operated by a rocking lever and pawl revolving a ratchet wheel is almost universally applied to passenger-hauling locomotives. This form of pump is equally suitable for flash-steam plants operating on reasonable pressures, the drive from the water feed pump, of say approximately $\frac{1}{6}$th engine speed, may be picked up to operate the lubricant pump. The operating pawl rotates a ratchet wheel of perhaps 40 teeth, providing one 'pumpful' of lubricant at every 240 engine revolutions.

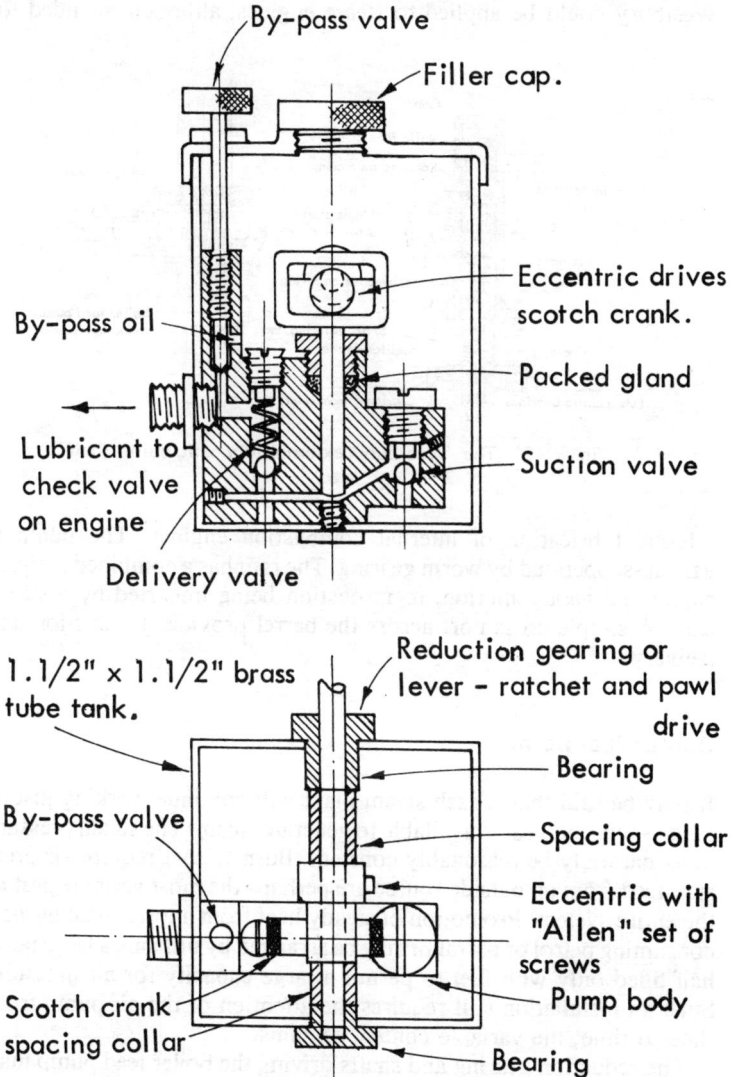

Fig. 7.14 The improved lubricant pump shown full-size

An alternative form of submerged lubricant pump, with valves and metering by-pass facilities is illustrated in *Fig. 7.14*. Like the oscillating pump, this may be operated by lever and pawls, driving the ratchet wheel and crankshaft.

A well tried lubricant pump as *Fig. 7.15*, designed by the late E. T. Westbury could be applied to steam engines, although intended for

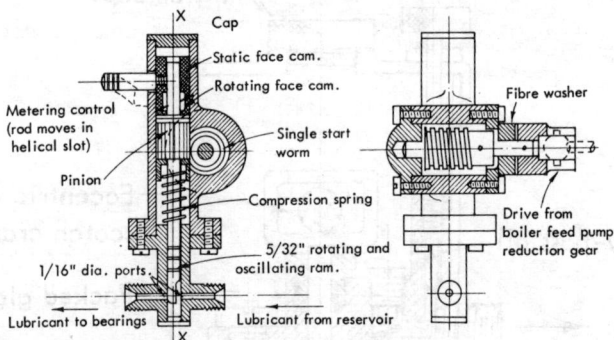

Fig. 7.15   The 'Westbury' external lubricant pump (half-size)

internal lubrication of internal combustion engines. The pump is valveless, operated by worm gearing. The ram has a combined reciprocating and rotary motion, reciprocation being imparted by a sleeve cam. A simple cross port across the barrel provides for suction and delivery.

Burner fuel pumps

It may be said that a flash-steam plant will continue working just as long as fuel and heat is available to generate steam. The steam pressure must naturally be reasonably constant. Burners that require air pressure feed from an outside source are perhaps the most variable part of the entire system. Promotion of steady heat from a torch-type burner, consuming petrol or parrafin, may be arranged by utilising a large tank, half filled only with fuel to permit a large capacity for air pressure. Such an installation still requires the attention of the air pump from time to time; the variable content remains.

The reduction gearing and shafts driving the boiler feed pump may be further utilised to operate other pumps, either through additional gearing or direct operating cranks. The principle of delivering liquid fuel to the burner in direct proportion to engine speed is obviously desirable, but some problems may be encountered in designing such a pump. Variation of the ram travel by adjustment of the operating crank and repositioning of the pump will provide for some flexibility. This arrangement is also employed for the boiler feed pump.

Several racing hydroplanes have incorporated fuel pumps, but all exponents have not fully exploited the system. Pumping petroleum has

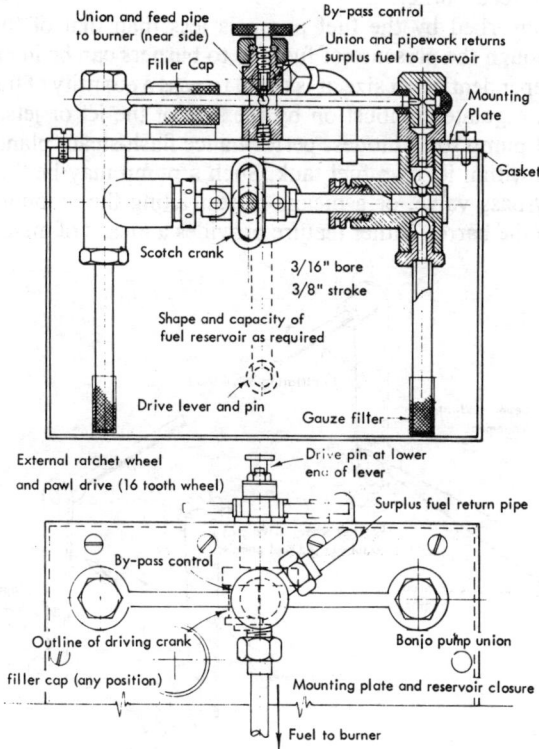

Fig. 7.16　Dual submerged fuel pump

promoted problems due to vaporising of the fuel within the pump and immediate pipelines, this vapour locking is, of course, very similar to the air locking of water feed pumps. A less volatile fuel such as paraffin reduces the problem. The fuel pump must be kept cool: not such an easy task in steam plants. A water-cooling jacket around the pump

barrel assists, and could in fact take the form of aluminium alloy block
into which both water and fuel pumps are built. When a fuel pump is
employed for boiler firing, the problem of starting up again has to be
considered. This involves another pump for hand operation, or at least
an auxiliary barrel and ram working in conjuction with the mechanical
pump valve box. Perhaps this complication has deterred some experimental power boat enthusiasts. Or the added weight and loss of engine
power to drive such pumps may outweigh the system in favour of the
air-pressure feed burner.

Power absorbed by the fuel pump is less than that of the water
pump; although the pressure of fuel fed to burners can be in excess of
150 *psi*. Dependent on jet size, it is usual to pass a quantity of fuel to the
burner and regulate combustion by the size of the jet or jets.

The fuel pump for a modest performance flash-steam plant can be
submerged within its own fuel tank. Such a pump may be fitted with
either a by-pass valve, or a means of restricting the amount of fuel
drawn into the barrel. Either feature provides a means of metering the

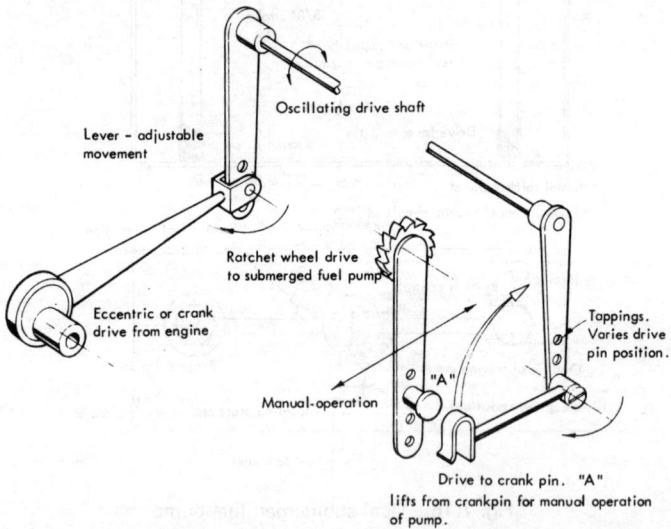

Fig. 7.17   Diagramatic plan of the remote drive for the fuel
pump

quanity of fuel passed to the burner. *Fig. 7.16* illustrates a design for a
pump. It will be seen that the operating mechanism is of the lever,
ratchet and pawl type, very similar to the lubricant pump. An advantage of this form of drive will be noted, as the linkage operating the

lever, may be detached from the engine drive (*Fig. 7.17*) to permit hand operation of the pump to start up the plant. In addition, the length of the drive-lever may be varied to pick up one two or three serrations of the ratchet wheel. This provides for operating the pump at differing speeds, it is most useful as the submerged form of pump, but unlike the isolated barrel and ram pump, cannot be varied to permit differing length of stroke and fuel delivery.

The submerged pump may be installed at some distance from the engine and other auxiliary components. This feature alone helps design and may eliminate heating of the fuel referred to above. Naturally the fuel surrounding the pump makes for cool working and lubrication of ratchet gear, when this is built into the fuel reservoir. *Fig. 7.17* indicates the arrangement for placing the fuel pump remote from the engine.

CHAPTER EIGHT

PLANT LAYOUT AND HULLS

Installation problems of flash-steam plants within the launch or ship model are not unlike other steam plants. Combustion, air for the burners, ventilation of hull interior and prevention of heat damage to hull and superstructure being the usual items receiving attention.

The burner is of prime importance to provide suitable conditions within the hull for continual operation. It is also essential that a good flow of air should reach the combustion zone. The nature of some prototypes, although attractive, following full-size steamships with one or more smoke stacks, produces problems. This is because superstructure details cannot provide the ventilation required for the combustion and reasonable temperature conditions. The type of vessel must

Plate 22 A complete flash plant by H. W. Saunders. Note the exhaust oil-trap and hand pump sited at the rear for convenience

be well considered before entering upon construction and installation of the steam plant. Obvious though this may appear to be, constructors find that ship models have to be altered after trials to permit the plant to operate efficiently. If sufficient ventilation cannot be made within the superstructure by functional cowl ventilators, open cabin doors and skylights, it may be possible to induce combustion air by a small electrical motor and disc fan. This may be fitted within the second smoke stack, or other convenient position inside the lift-off superstructure. The power supply for the fan motor is fitted in the hull and operates the motor when the superstructure is placed in position, locating with electrical contacts. Exhaust steam, discharged into the boiler uptake stack assists the flow of air and combustion (see following chapter), and adds realism to the model. It is usual to fit an exhaust trap in order to retain any oil carried over by the exhaust steam; indeed, the operation of steam-driven craft on some sailing waters makes this feature obligatory. In *Plate 22*, showing the complete flash steam plant of *Verulam*, by H. W. Saunders, the exhaust trap is clearly illustrated. A small tap at the lower end enables the trap to be drained when required. It will be noticed that the whole construction is neat and compact, and offers an excellent example of flash steam applied to a ship model.

As the launch hull is often only partly decked, ideal conditions are presented for operation of the burner. The diagram, *Fig. 8.1*, illustrates

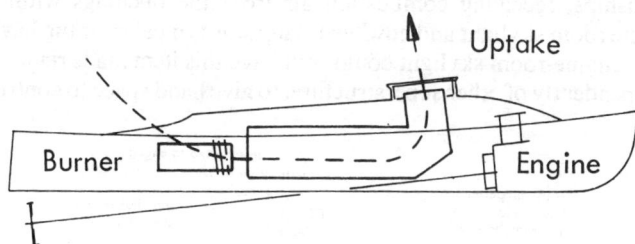

Fig. 8.1   A typical layout

a typical layout. The burner may be rigidly fitted and connected with the fuel tank, or the combined firing unit is made to lift out for starting up and then returned to the hull and held by spring clips or other devices.

Whatever machinery is installed, it is most important to obtain correct fore and aft trim and then load the vessel to the waterline as required by adding ballast; this consideration usually involves placing the rather heavy component—the drum type boiler in the most suitable position within the hull. As the flash boiler with its casing is by comparison, so much lighter, the approach to where the boiler shall be sited is more often governed by the uptake and position of the smoke stack. All other units of the plant may be built in to suit the boiler. Naturally, the location of the engine and power take off to couple with the propeller shaft or drive two or more propellers, must be considered. It is not possible to lay down the exact weights of steam plant and installation with ship models. Access to operate and maintain the plant is most important. *Fig. 8.2* indicates the disposition of the plant applied to a tug, this being a much-favoured model. It will be noted that fore and aft trim is provided by the feed water and fuel tanks built in.

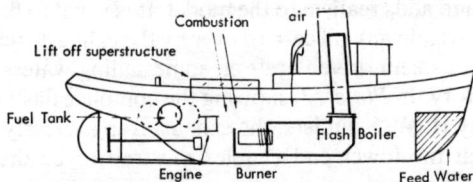

Fig. 8.2   Layout for a tug

The aft fuel tank is carried across the hull, and could well be made adjustable in variation of hull trim. The bow tank follows the contour of the hull and is the reservoir for boiler feed water. The burner is amidships, receiving combustion air from the openings within the engine room skylight and cowl ventilators. A typical river tug having a large engine-room skylight could well have this item made removable, independently of other superstructure, to give hand space to controls.

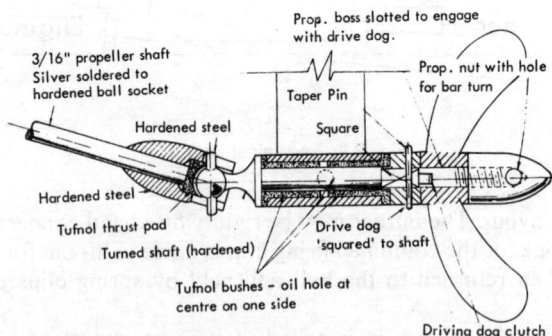

Fig. 8.3   Universal joint and propeller shaft

For the faster model launch; often a planing hull, the angle of the propeller shafting influences the performance of the craft. The shallow angle required dictates a forward position for the engine; often a universal joint and tail shaft still further reduces the angle through which the propeller operates. This form of skeg is indicated in *Fig. 8.3*, comprising the tail shaft revolving in a Tufnol bush and furnished with a 'dog' to drive the propeller. The design indicated is favoured by the authors, who have used the arrangement on all launch type craft. Just as the trim of the round-bilged hull of a model ship has to be carefully considered for correct flotation, the planing hull has to rise into the planing position, as well as return to the water and trim correctly at rest. 'Porpoising' of flat-bottom hulls is well known, and although offering something of interest to observers at the pondside, is unnecessary, and not so difficult to control if the arrangement of weight within the hull is carefully allocated. With the boiler amidships and the burner aft, the correct arrangement for trimming the hull may often be resolved by placing the fuel tank in the position required to trim the boat.

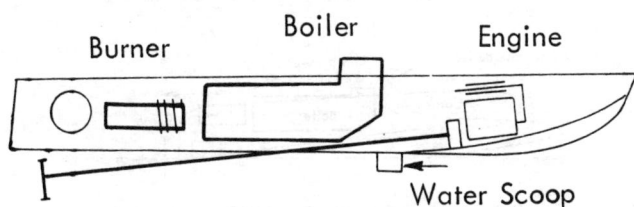

Fig. 8.4   Suggested layout of steam plant units

It is most convenient to provide for operation from the open cockpit. Boiler feed hand pump and burner controls, including air pressure gauge and cycle pump connection should be installed here, even if the fuel tank is remote. *Fig. 8.4* indicates the proposed location of the units forming the steam plant.

Flash steam has always been associated with racing hydroplanes,

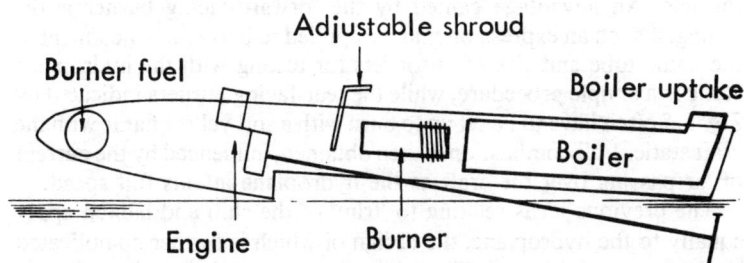

Fig. 8.5   Layout for early flash-steam hydroplane

the earlier craft following closely upon the launch type boat. The engine was placed well forward (*Fig. 8.5*), driving the propeller through a shallow shaft. Boilers were generally situated well aft, perhaps to ensure that any loose flame was wafted well clear of the hull. It was not uncommon for hydroplanes to 'cool off' during a run, this causing neat fuel to be discharged through the boiler to burn outside the casing. Many experiments were applied to keep the blowlamp alight. The fitting of shrouds and shields around the burner or burners either encouraged or prevented air from reaching the flame tube as the boat progressed against the current of air passing over the craft. *Fig. 8.5* shows the arrangement most exponents adopted, the simple adjustable deflector indicated could be a very important feature for any similar installation considered today.

The alternative location for the boiler and burners is that favoured for the launch *Fig. 8.6*. In this instance it will be noted that forward

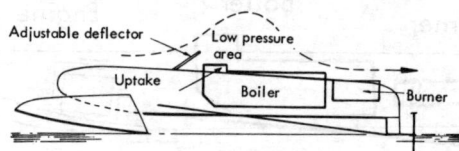

Fig. 8.6   Layout showing forward-facing burner

movement of the hydroplane does not scoop up air for combustion and possible cooling off of boiler and burner vaporiser.

Air flowing over the hydroplane tends to promote extraction of combustion gas from the boiler dependent on the shape and inclination of the adjustable deflector. The deflector may be hull-mounted, or in a simple sheet-metal extension of the boiler casing and uptake. It may equally be understood that such burners operate in a normal way, drawing combustion air as required, dependent upon gas issuing from the jets. An advantage gained by the forward-facing burner is the tuning, if such an expression may be applied to burners. Adjustment of the flame tube and size of jet or jets for testing with the hydroplane static is a simple procedure, while the rear-facing burners indicated by *Fig. 8.5* often have to be set up to burn with a soft yellow flame with the craft static. Full combustion is then obtained, influenced by the current of air passing over the craft as the hydroplane attains full speed.

The previous ideas relating to 'trim' of the ship and launch apply equally to the hydroplane, the design of which is further complicated by the nature of this craft. The additional problem before the designer, requires a hull form that will rapidly accelerate to lift the mass clear from the water into the planing or suspended position at high speed. It is also necessary to provide correct hull shape and sponsons to lower the hull when engine revolutions are reduced and deceleration of the craft takes place, and the craft again comes to float normally. This in itself is not so easy as it would at first appear. The hydroplane fitted with steam plant often comprises a box-shaped hull, sufficient to contain the propulsion machinery and perhaps a supply of boiler feed water. The size and weight of the hull, in the interest of high performance and conformity with class racing rules, is reduced to the minimum consistent with strength. The attached planes or sponsons, although adequate to maintain high performance with as little skin friction as possible, are generally inadequate for slower speeds. The steam-driven hydroplane, unlike the *ic* engine craft, does not come to rest quickly following operation of the stopping device. The gradual reduction of engine speed slows the craft which endeavours to operate as a displacement boat; the sponsons and tail shaft, efficient as a three-point hydroplane system, now have to function in a different way. Often the large diameter surface propeller (see below) lifts the stern of the craft, which in turn depresses the forward section.

It follows that the design of the hydroplane hull should provide adequate buoyancy in the forward section, together with hollow or ultra lightweight attached sponsons. The latter are usually arranged to provide an angle of incidence; that is to say, a planing angle with the surface of smooth water, of about 5–8 degrees. Shaping of the top of each sponson should be arranged to prevent lifting by aerodynamic action, as this is likely to capsize the craft when at speed. It would appear to be a simple matter to provide large sponsons to ensure adequate buoyancy; unfortunately, the extended underside would generate air lift and contribute to capsizing—it is all a question of compromise. *Fig. 8.7A* indicates the desired shape to provide negative

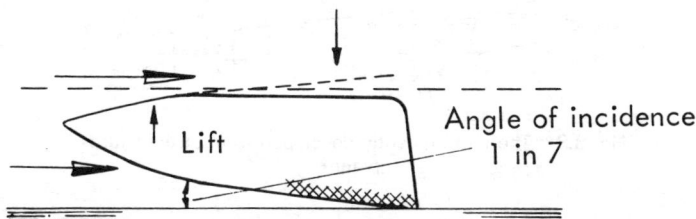

Fig. 8.7A   Shape of sponson to prevent too much air lift

air lift from the upper surface while *Fig. 8.7B* is the shape to be avoided.

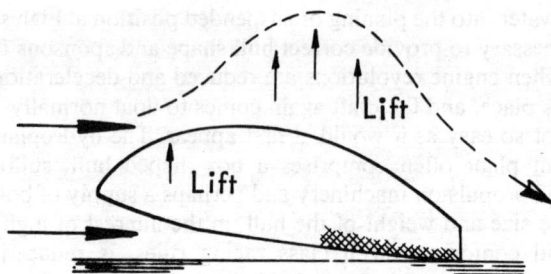

Fig. 8.7B   Incorrect shape of sponson

During the evolution of the steam-driven hydroplane constructors have been forced to accept the hull as a carrier for the machinery of propulsion, rather than conceive of it as a design in itself. Earlier craft with submerged propellers relied on a single step across the hull and wide aft plane to take the craft up, and maintain it in the planing attitude. Development of the three-point hydroplane and its application to flash steam introduced the surface propeller. Boats so fitted in full-size practice are referred to as 'Prop riders'. This implies that the hull is supported and rides partly upon the propeller boss. Here we may mention that some earlier lively boats fitted with submerged propellers showed a tendency for the stern to lift. Although experimental work generally involved holding the stern down to prevent propeller cavitation, the phenomenon was soon exploited to such an extent that a remarkable diversity in design occurred. We will return to propellers

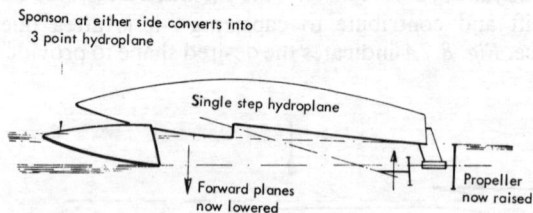

Fig. 8.8   The more recently developed form of the hydroplane

and propulsion later in this chapter, but must now relate the new system as applied to the flash-steam hydroplane.

It is logical that the single-step hydroplane should now be adapted and furnished with two forward planes or buoyant sponsons in replacement of the step. The sponsons are somewhat lower than the traditional step, while the depth at which the propeller operates is raised (*Fig. 8.8*). This is the basic form of hydroplane operating today, with many variations used for racing craft fitted with internal combustion engines. For the flash steam plant, further complications arise. We have already set out the buoyancy required, acceleration and problems of slowing the craft after operation of the stopping device. As the high speed craft may be partly airborne, the very action of the hydroplane is in jeopardy if water has to be scooped from the lake for generation of steam. It has been common practice to scoop water into a small reservoir and filter, from which the feed pump draws its supply. Recent development indicates that the water on which power boats operate is far from suitable for the pump and engine. By elimination of the scoop, a little more resistance is removed in the interest of speed. It may be considered that carrying the feed water, with its associated pumping, offsets any gain by loss of drag. This is not borne out in practice. If the design for a hydroplane relies on scooping water from the lake, the scoop should be fitted on the inner side of the inner sponson (*Fig. 8.9*).

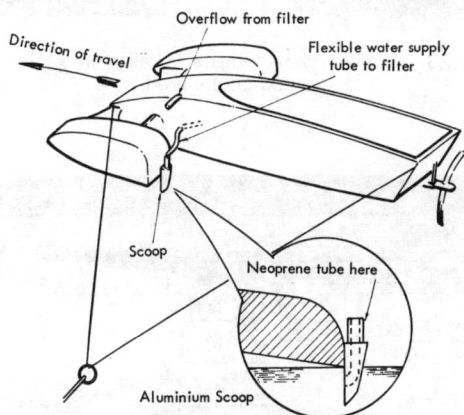

Fig. 8.9   Arrangements for the water scoop

The mouth of the scoop should face the direction the hydroplane takes while running on the circular course. The inner sponson can often be persuaded to stay on the water by 'banking' the craft a little, by the bridle arrangement.

The traditional forward engine position so necessary with submerged propellers, may now be abandoned, and the engine placed at the rear of the craft for driving the propeller through a short shaft supported by flexible or universal couplings. The late B. Pilliner's

Plate 23　Single-cylinder engine installed in an inclined
position (J. Bamford)

Plate 24　A twin-cylinder poppet valve engine installed in
an A-class hydroplane (S. Poyser)

hydroplane was most successful, driving the propeller through a bearing in the transom of the boat. A rather daring break with tradition enabled Mr. Pilliner to build the engine across the transom of the hull, the flywheel carried outboard, incorporating the propeller blades. Effective propeller diameters in the order of 6 inches were used (*Fig. 8.10*). Several exponents still prefer the 'forward' engine as depicted in *Plates 23* and *24*.

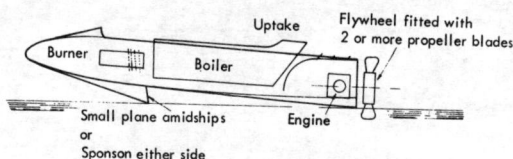

Fig. 8.10   Rear-mounted engine

Other experimental work with aft-mounted engines driving short propellers shafts either through the transom or to brackets and small planes has been successfully carried out by Messrs. Jutton and Bamford. *Fig. 8.11* indicates the hull, which may be either two or three

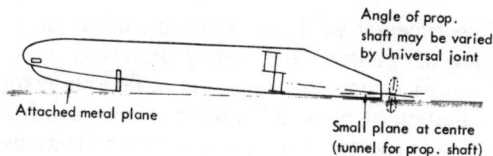

Fig. 8.11   Experimental design with aft engine

point suspension, having either a central plane beneath or a small sponson at either side.

The universal joint, useful though it is to provide for alteration within drive shafts is a potential waste of power; its mere presence is yet another component that may fail by mechanical breakdown. Very high stresses are imposed upon these small parts, often subjected to shock loading and yet expected to transmit the full power of the engine, which may approach 3 H.P. *If* the design of the hydroplane can be resolved with a simple propeller shaft and without such universal joints, then this is good design in itself (*Plate 25*).

Despite modern technology in reinforced glass-plastic hulls, the most favoured lightweight materials are plywood and selected hardwoods. Strength is beyond doubt when associated with modern resin

Plate 25 'Straight-through' prop shaft on *T.N.T.* (F. Jutton). The 'butane' tank actually contained paraffin/petrol

adhesives. Experimental work, so necessary in developing a hydroplane, variation of sponsons and adding bearers is accomplished by cleaning off varnish or paint to expose the surface ready for a new glue joint. One limitation of plywood is the possible absorbtion of water and oil, resulting in increased weight. Several applications of paint or varnish to resist penetration are necessary, but over-enthusiasm here will add perhaps more weight than would result by absorption. Flat pieces of plywood in themselves are rather flexible and in larger pieces too flexible for hydroplane construction. The designer has either to build in more frame and stiffening fillets, or produce a hull form involving curved surfaces, naturally the latter is favoured, not only for the more pleasant eye appeal but for the weight-saving quality. Plywood behaves like any flat sheet; it can only accept curvature in one direction. Compound curving is possible by building a simple mould of the desired form and resorting to the process favoured in the boat building industry known as cold-moulded plywood. This form of construction, using two skins of 1 *mm* plywood, laid crossing and bonded to each other at approximately 90 degrees results in a very strong structure.

Installation of the internal combustion engine to the hydroplane offers the designer considerable scope in formation of an almost closed hull, the decking contributing considerably to the strength of the hydroplane frame. The flash steamer, crammed with machinery, generally prohibits closure of the hull other than the forward section. Because of the array of components that have to be built-in and supported with adequate fastenings, in most instances the hull can be no more than a container. The sides of the hull may be stabilised by bonding them through parts of the plant. The width of the boiler case is often sufficient to incorporate lugs or seatings which, with an insulating material, act as a cross-stay at the upper portion of the hull: see *Fig. 8.12*. The boiler casing must in itself be sufficiently rigid to perform the function of the stay.

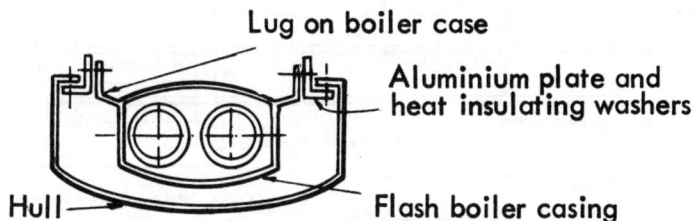

Lug on boiler case

Aluminium plate and heat insulating washers

Hull

Flash boiler casing

Fig. 8.12   The boiler plant acts as a cross-stay. The flash boiler casing is supended from the hull coaming

It is our hope that readers will feel sufficiently enthused to pursue speed with the flash-steam plant and require a design to follow in building the hull. Some digression is now called for in explanation of rules relating to racing craft as recognised by the Model Power Boat Association.

Several classes of racing craft may be built for competition, and where as in earlier days the largest engines produced the highest speed, this is not so today. The small-capacity internal combustion engine has reached a stage of high efficiency, propelling craft at speeds exceeding those fitted with larger capacity engines. Although the number of flash steamers operating are very limited, it is found that the size of engine and craft is not the dominating factor in pursuing high speed. Construction of steam hydroplanes in the smaller classes is not justified. Craft should be constructed to align with two of the present racing classes, 'A' and 'B'. The difference between these is merely a weight limit. The larger 'A'-class hydroplane having a maximum weight limitation of 16 lbs., while the smaller 'B'-class craft must not exceed 8 lbs. Fuel, water and lubricant required for each class is not included in the weight.

As a typical racing plant is described elsewhere, it is only necessary to indicate a simple hull in which the plant may be used. *Fig. 8.13* is not dimensioned as the outline and relevant information offered may be applied to either an 'A'-class or 'B'-class boat by variation of dimensions. As a guide, the 'A'-class hull would measure about 3 feet and have a width over the hollow sponsons of 12–15 inches. Curvature of the plywood skin forming the floor provides for rigidity. The hollow

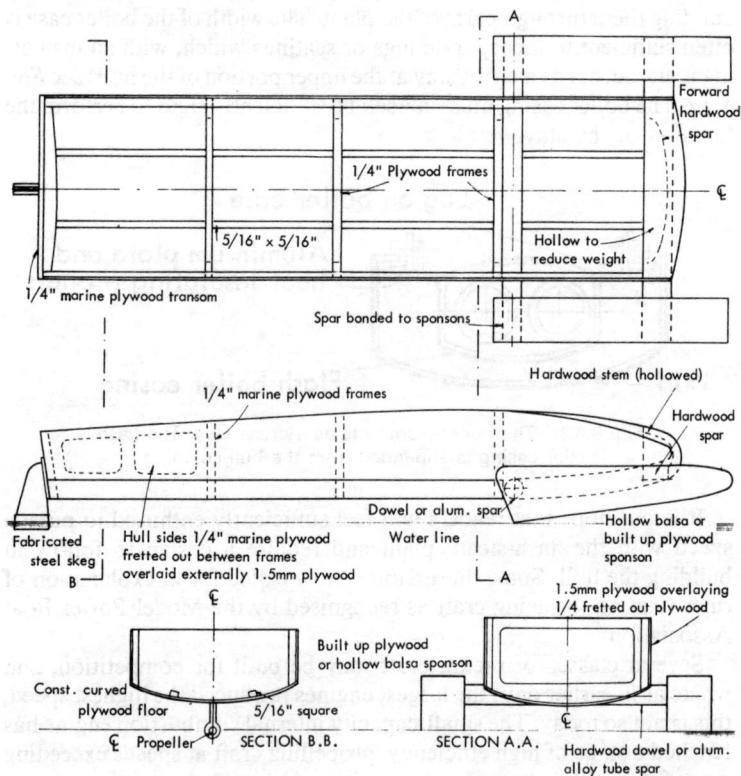

Forward hardwood spar

1/4" Plywood frames

5/16" x 5/16"

Hollow to reduce weight

1/4" marine plywood transom

Spar bonded to sponsons

Hardwood stem (hollowed)

Hardwood spar

1/4" marine plywood frames

Dowel or alum. spar

Water line

Hollow balsa or built up plywood sponson

Fabricated steel skeg

Hull sides 1/4" marine plywood fretted out between frames overlaid externally 1.5mm plywood

1.5mm plywood overlaying 1/4 fretted out plywood

B

A

Built up plywood or hollow balsa sponson

Const. curved plywood floor

5/16" square

Propeller

SECTION B.B.

SECTION A.A.

Hardwood dowel or alum. alloy tube spar

Fig. 8.13   The outline of a typical three-point hydroplane

sponsons are supported and attached to the hull by two spars, one of which is *in situ* with the stem framing. An experimental hull of novel design by Frank Jutton is shown in *Plate 26*.

Whatever system is employed to power a water-borne craft, the screw propeller is the accepted device acting in or on the water. Regrettably, it is not possible to specify propeller sizes, pitches or blade shapes. Established designers of propellers used in full-size marine practice are better equipped, by reason of experience and sales to boat builders. All craft undergo trials by running over measured distances, with and against tide, varying winds and seas. Often a change of propeller or propellers is made and all data collected and fed back to engine and propeller builders to extend the bank of technology.

Model power boating is not unlike full-sized practice, other than the fact that there is little knowledge to fall back on in selecting a propeller to begin the experimental stage of propeller development. Model

Plate 26   Experimental hull by F. Jutton. In the event the enclosed plant suffered from an insufficient air supply

power boating certainly involves running 'trials' to determine optimum results, and indeed it may be understood that the trials of the hydroplane never reach finality.

Each constructor will have to build and experiment with propellers; Model trade suppliers have yet to offer larger propellers for even the lower-performance steam plant. Fabrication in brass or steel with silver solder jointing is the accepted procedure; a simple jig assisting to hold the blades into the slotted boss during this operation.

For ship models good results are usually obtained from those having a pitch ratio of 1, i.e. a 3-inch diameter and 3-inch pitch propeller. The lighter launch-type craft, often with higher engine revolutions, accepts a propeller with slightly higher pitch ratio of about $1\frac{1}{2}$. Blades may vary from one propeller to another but are often rounded or oval in form. The hydroplane with high-speed engine may well revolve a propeller of $2\frac{1}{2}$ or 3 pitch ratio—blade shape is generally narrower, and of wedge or cleaver section. As a guide to pitch, an angle of 40 degrees at $\frac{2}{3}$ diameter of the propeller produces pitch amounting to approximately twice diameter.

For propellers operating fully submerged, blade thickness should be kept to the minimum compatible with strength—thus built up steel is favoured rather than brass or gunmetal castings. The surface propeller has to withstand considerable shock, it is therefore necessary to make such propellers of high grade steel. Although not universally acc- laimed, the cleaver-shaped blade is often incorporated, the leading edge is sharp while the trailing edge is left thick—a wedge section.

It is now necessary to quote another part of the M.P.B.A. competi- tion rules relating to hydroplanes: "All boats must be fitted with a switch which can be operated when the boat is running." This is commonsense and requires the fitting of yet another component. The stopping device must be efficient in action and capable of being tripped as the craft laps the circular course. Whatever the form and motion of the stopping device, the operating lever must project 4 inches or more beyond any other projections of the craft.

One way to slow and stop the flash plant is to open a boiler release so starving the engine of its energy. This arrangement works well in steam plants that incorporate burners receiving fuel from the engine-operated pump. As the engine speed falls, the quantity of fuel passed to the burners diminshes until all heating of the boiler ceases.

Alternative arrangements must be provided for craft fitted with air- pressure fed burners (*Fig. 8.14*). It is not sufficient to release the

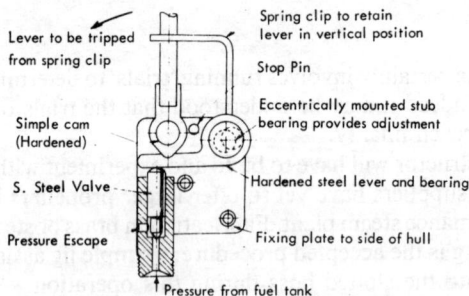

Fig. 8.14   Stopping device (handed to suit direction of hull travel)

pressure of steam, hoping that the craft may be retrieved before damage results to the boiler tubing. A simple stopping system would require only the fitting of a lever-operated valve to release the fuel pressure. This would permit the burners to die down and out, even- tually bringing the hydroplane to rest. It is more usual to provide valves that may be operated by the lever in releasing both air and steam. This brings the craft to rest in a shorter period of time than would otherwise result merely by releasing air pressure.

CHAPTER NINE

A SIMPLE FLASH-STEAM PLANT

Some flash plants seen in non-racing craft have employed engines that are not really suitable for easy or neat adaption. Although bedplates have been used supporting gear shafts, pumps and other bits and pieces, they often do not look very nice, are difficult to keep clean and anything but compact.

In making the above statement we do not intend to disparage the efforts of constructors who have produced flash plants that have worked well—often with very limited equipment indeed.

The following description is intended for those who have reasonable facilities and wish to make a compact plant using a ready-made engine suitable for flash steam. The engine chosen is the well-known Stuart 'Sun' which is still available as sets of castings or as a finished engine. A piston valve is standard, but a slide valve chest is also available. The former is perhaps the most suitable, but slide valves are in order if cast iron is used instead of gunmetal or bronze.

A full description of machining methods together with dimensioned diagrams is considered appropriate as it is hoped that this will assist the less experienced model-maker to 'do the necessary'.

The 'Sun' engine has a sturdy crankshaft supported in generous bearings and an enclosed crankcase. By variation of one bearing the engine may be equipped with a secondary shaft driven through reduction gearing to operate water feed and oil pumps. *Fig. 9.1 (a, b, c* and *d*), indicates the adaption. The box-shaped crankcase is very suitable for fastening brackets to support boiler feed and oil pumps.

A little work with a file is required to clean up the exterior of the crankcase casting, or a simple milling operation may be carried out to machine these faces.

Although the worm gearing might be unsuitable for a racing job, it may be considered adequate here; it is compact and can be obtained from a well-known model supplier's (Bond's). These gears consist of a steel worm and a hard brass wheel giving a ratio of 6 : 1. Other worm gears of similar ratio and external size may be used if shaft centres, etc. are modified to suit. To run well it is important that the worm and

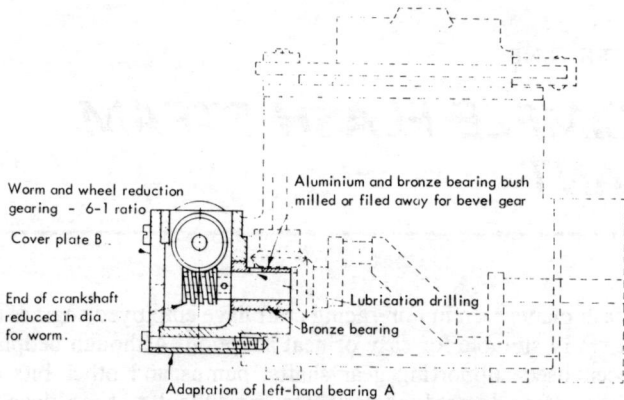

Fig. 9.1a   General arrangement of reduction gearing and adaption of bearing

wheel mesh without backlash, and end float must be controlled on the worm shaft.

If the standard 'Sun' crankshaft is to be machined from scratch, it is a simple matter to provide for the fitting of the worm, but if a finished

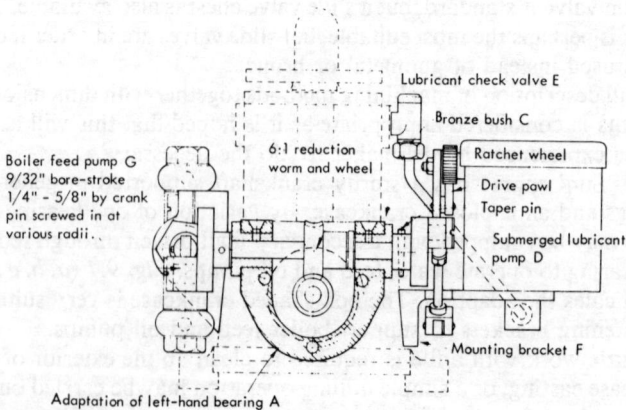

Fig. 9.1b   General arrangement of reduction gearing and pump drive

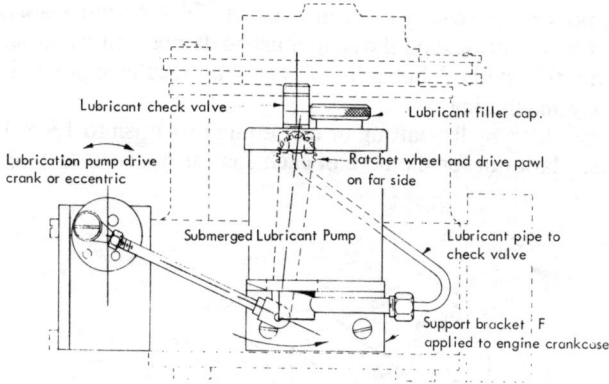

Fig. 9.1c   General arrangement—lubricant pump drive and
mounting

engine is to be adapted, the crankshaft will have to be set up in the lathe
to reduce the diameter a little to fit the worm. Later the worm is pinned
in position.

A block of aluminium alloy should be used for the new component

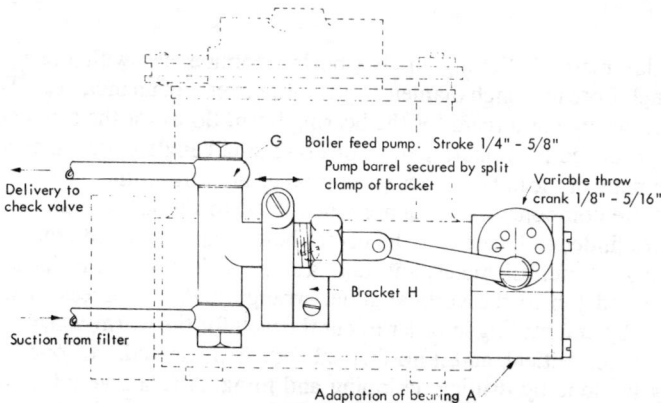

Fig. 9.1d   General arrangement—boiler feed pump—drive
and mounting

at A (*Fig. 9.2*). Stresses on this part should not be great, and if no alloy
bar is available, a simple cast block would serve. Simple, but effective,
castings can be easily produced by melting some scrap aluminium in a

F

ladle and pouring in a cavity made in a bucket of damp sand. Care must be taken when pouring, and the sand must be damped only a little. We do not extol this method, but at least it gives fair results in practice and allows easy machining.

Face the block by fly-cutting or end-milling to finish to $1\frac{9}{16} \times 1\frac{5}{8} \times 1\frac{1}{2}$ inches. Mark out for the bore of each shaft using surface gauge or

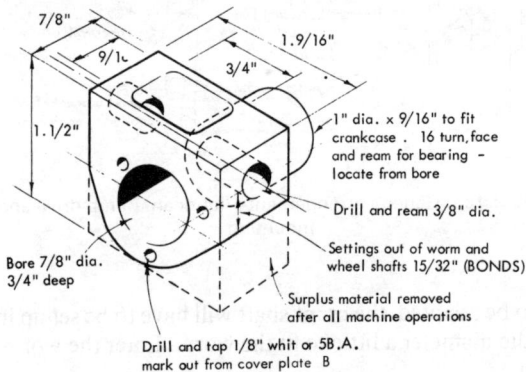

Fig. 9.2   Detail of new component in adaption of bearing A

similar method. Setting up for boring commences with the worm tunnel. Bore to $\frac{7}{8}$-inch diameter and a little over $\frac{3}{4}$ of an inch deep. Drill right through and ream for the bearing bush. Bores for the cross-shaft are produced by mounting the block on a small angle-plate attached to face-plate of the lathe. A facing cut is taken after the initial operation so that the component sits square when fixed to the angle plate. Use a centre finder to locate at lathe centre height and drill, and ream right through $\frac{3}{8}$-inch diameter, and take a light facing cut. Transfer angle plate and job to the vertical slide, arranging at 90 degrees, remove metal by end-milling in order to cut the well for the worm-wheel—the size is not critical, and if no vertical slide is possessed, this operation may be done by drilling, chipping and filing. Alternatively the work may be packed up on the lathe cross-slide and some metal removed by an end mill or slot drill.

Mount the work on a mandrel to face the far side of the reamed hole, supporting the mandrel end by the tailstock centre. Use a stub mandrel to turn the spigot that has to locate in the crankcase in replacement of the original bearing. This diameter will require a flat, machined or filed after fitting the bush in order to clear the bevel gears of the shaft driving the valve. Do not overlook the small drilling provided to lubricate the journal. The bearing bushes are press fitted, then reamed in position.

A simple cover plate—B—is turned up and drilled with three No. 40 holes, this cover can then serve as a jig for drilling the holes through component A and the crankcase can be 'spotted' so that all holes are in alignment. Crankcase holes are tapped 5 B.A. or $\frac{1}{8}$-inch Whitworth.

Examination of a standard engine will show that the existing tapped holes for the bearing cannot be used as their location is unsuitable. The worm wheel is secured to a piece of round $\frac{3}{16}$-inch diameter rod with a taper pin, or 6 B.A. Allen screw, dimpled into the rod. Make spacer collars to locate the wheel centrally on the worm.

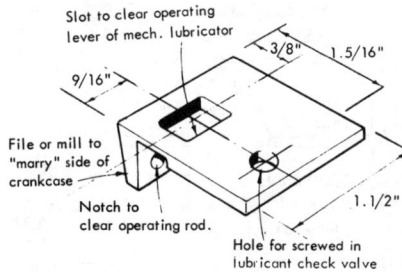

Fig. 9.3   Detail of submerged lubricant pump, supporting bracket F

Boiler feed pump and lubricator are operated by connecting rods driven from disc cranks having screwed-in crankpins, these enable the throw to be changed in order to vary the stroke. Crank discs are turned up, press-fitted to the shaft and fixed with a small taper pin. Small Allen screws may be used here, but must be fully tight, and the shaft dimpled to receive them.

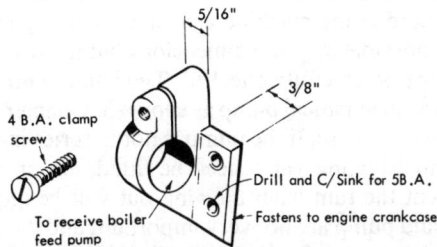

Fig. 9.4   Detail of mounting bracket H for feed pump

With the above assembly complete the engine is ready to receive the pumps. *Figs. 9.1b, 9.3* and *9.4* will indicate the attachment of brackets H and F. The sides of the crankcase are not thick enough for secure tappings, attachment should thus be made with screws having nuts inside the crankcase. Use large washers to spread the load on the inside. Washers for the lubricator bracket F will have to be tapered and can be made by filing a suitably drilled piece of aluminium alloy.

The boiler feed pump G (*Fig. 9.5*) is built-up from brass round and

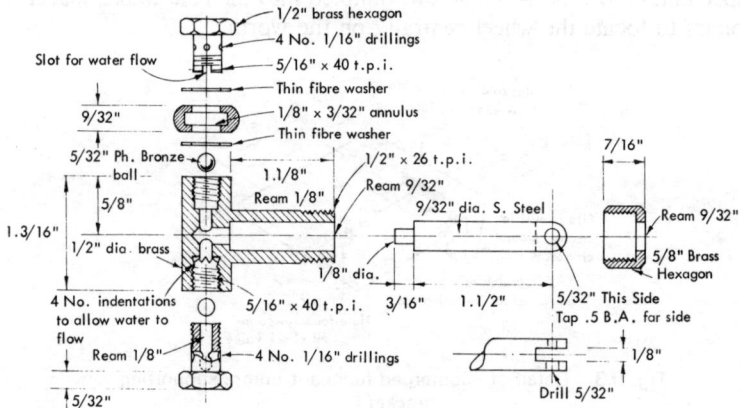

Fig. 9.5   Boiler feed pump G

hexagonal bar. The joint between barrel and valve chamber is made by filing the barrel to suit, and then silver soldering. Valves are $\frac{5}{32}$-inch bronze or stainless steel balls seating on $\frac{1}{8}$-inch reamed holes. The seats must be finished flat with a D-bit, not just drilled. Valve lift is controlled by castellation of the delivery union fitting and similar indentations made in the pump body immediately above the suction valve. A small dental burr held in a hand drill may be used for this operation, or hold in the machine vice on the drilling machine. Minor variations are possible in pump dimensions but the restricted lift of the ball valves must be carefully checked. The hand pump may be incorporated with the mechanical pump as shown in Chapter 7 (*Fig. 7.9*), or it may be sited elsewhere. If the pump is not in series in the main water supply line, no isolating valve need be fitted, and in this instance a catch to prevent the ram from blowing out will be required. Dimensions of the hand pump are not very important, and it can be similar to the mechanical pump but fitted with a suitable lever for hand operation.

Due consideration should be given to external lubrication, although the large crankcase may be charged with lubricant before each operating period. Action of the big-end and bevel gears will splash plenty of oil around. 'Going round with the oil-can' on the minor bearings may be sufficient for short runs, but some constructors may prefer something better. A small header tank and oil gallery with drip feeds will be adequate for pump shaft bearings and valve crankpin, but if really effective big-end and main bearing lubrication is wanted, a pressure feed system using drillings through the crankshaft is required. This would be advisable if a fast-planing boat is envisaged, or if the engine is required for long periods of heavy duty. Pressure feed can be from a sealed container lightly pressurised by a cycle pump—a separate oil pump is not a necessity, but an oil control valve would be needed.

Internal lubrication for the piston valve and cylinders is fed through the small check valve shown in *Fig. 9.6*. This is screwed into the engine

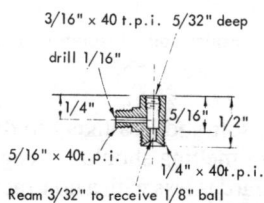

Fig. 9.6   Lubricant check valve E

instead of the traditional displacement lubricator. Oil is supplied by a submerged oscillating pump D (*Fig. 9.1b* and *9.1c*. These pumps are well-known in small locomotive construction, and they are simple and reliable. A standard 32-teeth ratchet wheel will be suitable if the pump is made with an $\frac{1}{8}$ inch diameter ram and $\frac{1}{4}$ in stroke. If preferred other types of pump could be adapted, but the ratchet drive has certain advantages. For example, the arrangement suggested gives a reduction of 192 : 1 but by altering the position of the crank throw or the pivot position the arm will swing through a greater angle and move two or more teeth, thus increasing the amount of oil supplied.

The flash boiler

About sixteen feet of $\frac{1}{4}$ inch diameter tube is arranged in a single coil so that it will fit into an insulated casing $10\frac{1}{2}$ inches long. The general arrangement is illustrated in *Fig. 9.7*. Heat is supplied by a petrol blowlamp having a flame tube of $1\frac{1}{2}$-inch diameter. The metal casing is

extended partly to surround the burner, thus protecting the hull and
allowing the lamp to be started up without removal from the boat.

We have discussed in Chapter 5 the merits of various materials.
Stainless steel is the first choice. For a plant of moderate performance
only, copper is perhaps next on the list, as it has a much longer life than

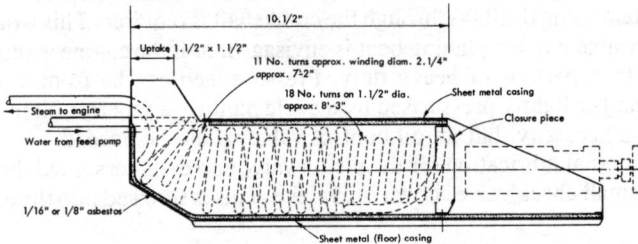

Fig. 9.7   Boiler coil—insulation and casing

mild steel, and scale is soft and less likely to damage the engine. The
biggest danger is that of melting should the engine stop with the lamp
burning—even this hazard is less with a non-racing plant as the lamp is
not usually working flat-out. If mild steel tube is used some means of
detaching the steam pipe from the engine is necessary, so that the
boiler can be 'blown down' before each period of running. A steel
union at the engine will suffice for this, if of suitable size.

It is an advantage if the boiler can be wound from a single length, but
if this is not possible a welded or sif-bronzed joint is permissible. Use a
sleeve of the same material to enclose the two ends before making the
joint. Silver solder is not really suitable here as there is always a risk of
the joint failing should the temperature rise much above 700 degrees C
(less for 'Easy-flow' pattern hard solders).

Anneal the tube, if necessary, and wind on a $1\frac{3}{8}$-inch diameter
mandrel. This mandrel may be of wood, steel water barrel or anything
of suitable rigidity. Start the winding in the vice—once started it will
wrap round nicely if of reasonable thickness. You can use 20 gauge or
even 18 gauge tube, if weight is no problem, and it will coil up much
easier than the thinner stuff.

Before coiling leave about 9 inches for the steam pipe and when 18
turns have been made the tube must be wrapped over the coil just
made. A piece of rolled strip or tube may be slid over the first winding,
to enable this to be done without it disturbing the tube coil. Pull the
split tube along as the second wind progresses; keep the first mandrel
in position during these operations. A tail of about 6 inches is left to

receive the check valve to be detailed later. When making the coil, try and arrange joints to be out of direct heat if brazing has been used.

Depending on the finished result the coil will now require some manipulation to fit the casing, and to prise apart or compress individual coils so that they are fully exposed to the flame. The flue end of the casing is upswept and the coil should follow this shape. It is doubtful if the finished coil will exactly resemble that shown in *Fig. 9.7*, but some irregularity is advantageous in collecting heat, providing the flame is not baffled unduly. A simple rectangular boiler casing is suggested. This can be folded up from thin tinplate, steel or more exotic material. Blue planished steel or stainless steel looks very attractive, and the latter is practically everlasting. However, availability will no doubt dictate the choice.

Fig. 9.8 indicates the construction of the casing, which consists of

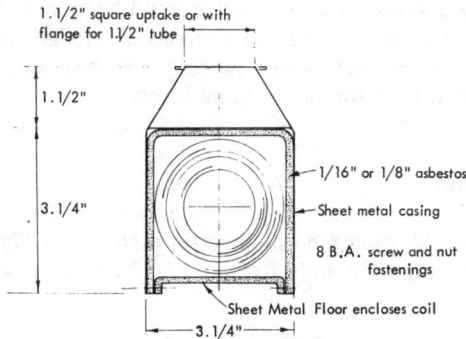

Fig. 9.8   Section of the coil insulation and casing ($\frac{1}{4}$ full-size)

an inclined floor covered with an inverted box. Further closure is provided by the flue construction and burner closure piece. The floor of the casing is inclined at the flue end to connect with the uptake. For a launch having simple superstructure, the tapered uptake can be sited below a simple funnel or ventilator, while a ship model could have a matching flanged joint at the top of the casing.

Whenever weight permits, the maximum amount of asbestos should be used and in any case an insulation lining must be placed on the inside of the casing. If the asbestos is wetted slightly it will fold without breaking down. Press into the casing leaving some asbestos projecting, a little heat will dry out the asbestos and 'solidify'. Now remove, so that the aperture may be cut for the uptake. When replacing, a few dots of gasket jointing material will hold the asbestos in position while small screws and nuts are used to clamp the metalwork and asbestos together. There is no need to fix the boiler coil to the casing, the inclined shape together with the steam and water connections holds all secure. For ship models, additional asbestos may be required around the uptake, and an elbow could be incorporated, so that exhaust steam may be discharged through the funnel to assist combustion.

The burner end of the casing need not be fitted with the closure piece shown, but it may help in fastening the asbestos lining and prevent delamination. Also it stiffens up the rear end of the casing. Some constructors may consider alternative fastenings, the pop rivet being the most obvious. Screws and nuts are, however, the most suitable for replacement of the asbestos, which will probably not last more than two seasons.

The casing may be finished by trimming any surplus asbestos or sheet metal. We suggest that the boiler could be mounted on aluminium alloy angle, extended fore and aft to accommodate the remainder of the plant. Two neat metal straps passing over the casing and secured to the angle will hold the boiler quite firmly.

The blowlamp

A torch pattern blowlamp burning petrol is shown in *Fig. 9.9*, which includes constructional details. Refer to *Fig. 9.7* for arrangement of the

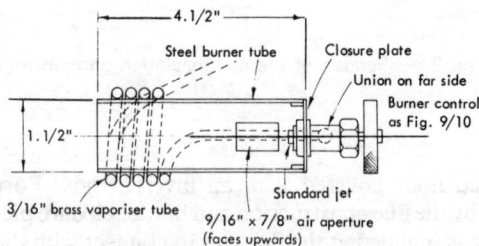

Fig. 9.9   Torch burner ($\frac{1}{4}$ full-size)

burner in the casing. It would be possible to use propane gas burners in place of the blowlamp providing that the boat could accommodate the gas cylinder. Such burners have been discussed in Chapter 5, and there is no doubt that these burners are now used quite successfully in many steam-powered craft.

The blowlamp has an auxiliary control valve behind the jet, and although not an adjustable jet in the accepted term, its function is similar, giving delicate control of the flame. A main control built into the fuel tank is also required to provide on/off burner control. The variable jet should be adjusted for best combustion, then left at this setting.

Air pressure applied by cycle pump is usually limited to about 60 *psi,* and for some purposes a standard jet as used in painters' blow-lamps may not pass enough gas. It follows that some enlargement may well be necessary. Some tests should be carried out with the lamp placed in the boiler within the boat. Experiment with differing fuel tank pressures, in conjunction with one or two different jets, will soon establish correct combustion. Standard jets usually vary between ·009 inch—·012 inch, and enlargement to about 018-inch diameter may be necessary.

Measurement of such tiny holes is possible by coating a very small sewing needle with engineers' blue and offering up to the jets. A bright ring will appear when the needle is twisted. This enables a micrometer to be used across the ring and a measurement taken. To enlarge the jet, convert a sewing needle into a broach by honing with a carborundum slip—the flats so formed will act as cutting edges. Hold in a pin vice and revolve in the jet from the back until the required size is obtained.

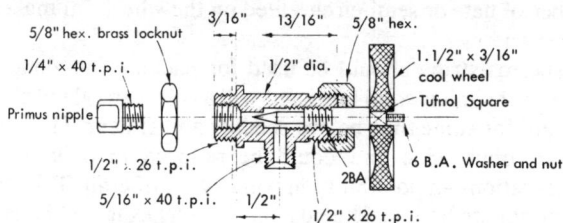

Fig. 9.10   Burner control

In order to reach all parts of the boiler, the flame has to be about 7 inches long, and in daylight should be bluish and almost invisible, burning steadily at all settings of the control. The adjustable jet is illustrated in *Fig. 9.10*. With the valve body turned from $\frac{5}{8}$-inch diameter brass rod reduce to $\frac{1}{2}$-inch diameter for 1 inch, face, centre, and drill 1 $\frac{3}{8}$ of an inch depth (No. 43). Now open out to $\frac{3}{4}$ inch deep with a No. 26 drill following with a D-bit or flat-bottom drill. Tap 2B.A. for $\frac{9}{16}$ of an inch, removing the first two threads with a clearing drill. Now thread $\frac{1}{2}$-inch × 26 T.P.I. for $\frac{1}{4}$ of an inch to receive the gland nut; reverse in the chuck to machine the jet housing. Standard jets have a thread very near to the $\frac{1}{4}$-inch × 40 M.E. thread—any difference is too small to bother about on such a short thread. Face up, and open the No. 43 drilling to $\frac{7}{32}$ inch diameter then tap $\frac{1}{4}$-inch × 40 for $\frac{1}{4}$ inch deep.

The end of this thread must be provided with a chamfer to suit that on the jet seating—a large centre drill will serve for this operation. Turn the outside to $\frac{1}{2}$-inch diameter and thread 26 T.P.I. for the locknut. Clean the No. 43 drilling from the gland end using any suitable reamer, D-bit or sharp drill.

The vaporiser connection requires a piece of $\frac{5}{16}$-inch diameter round brass drilled right through and countersunk for the olive. Reduce a short length to $\frac{1}{4}$-inch diameter to press into a matching hole drilled halfway along the valve body. Make a silver-soldered joint— preferably with Grade 1 or 2 strip rather than Easy-flow, as a higher melting point is desirable for safety reasons.

Use stainless steel or bronze for the valve spindle; part is reduced to $\frac{1}{8}$-inch diameter from $\frac{3}{16}$-inch finishing with a shallow point. Part is threaded 2 B.A. with the control end also reduced to suit the control wheel. A piece of $\frac{1}{8}$-inch thick Tufnol is suggested for the wheel as it will keep quite cool and burnt fingers will be avoided. To fix the wheel, square the end of the spindle and drill and file a hole centrally in the Tufnol. A small nut and washer retains the wheel in position. A number of flats or semi-circles filed on the wheel rim makes operation much easier.

Asbestos string should be used for packing the gland and a little graphite or similar solid lubricant will assist—the gland should remain gas-tight for some time before needing adjustment. This component is on the robust side and could be reduced in bulk if weight is a consideration—a possibility in some types of craft. The gland nut and anchor nut are both machined from $\frac{3}{8}$-inch hexagonal brass. Reference to the diagram indicates the attachment of the valve to burner tube. The gas issuing from the jet should pass cleanly down the centre of the flame tube, thus the end closure of the tube must be fitted square to axis.

True up the ends of a piece of light gauge steel tube $1\frac{1}{2}$ inches in diameter. If no tube of suitable gauge is available roll up from sheet, riveting or clenching the seam. An end closure may be turned from a piece of any suitable scrap or a steel disc brazed in position. A central hole is made $\frac{1}{2}$-inch in diameter to fit the threaded body of the valve.

The size of air aperture shown on the side of the flame tube should give good combustion, but can be easily enlarged if more air is required. An alternative is to make the aperture in the form of louvres, as mentioned in the following chapter.

The vaporiser has now to be made and fitted. We recommend $\frac{3}{16}$-inch diameter brass tubing in preference to copper. The latter tends to produce fine scale which may block the jet; stainless or mild steel can also be used. A length of 3 feet will be quite sufficient. Wind 4–5 turns on a $1\frac{1}{2}$ inch diameter former held in the vice. Place on the flame tube and arrange one end to line up with the jet valve. The pipe end is flared to suit the olive, remembering to slip on the union nut first. A similar union should be fitted at the control valve on the fuel tank.

Size and location of the fuel tank is mostly dictated by the type of boat, therefore no specific dimensions will be given. Although soft solder has been used for jointing fuel tanks, silver solder is much stronger and safer. Brass or copper is the best material, and 20 S.W.G. ample except for very large diameters—perhaps over three inches diameter ought to be 18 S.W.G. Remember that only half the capacity should contain fuel so that if plant is required to run for long periods— as might be likely with some ship models—the tank will need to be fairly large. A reliable pressure gauge is a very useful fitting on the fuel tank, assisting in regular and steady output of heat. Larger tanks can have an air pump incorporated if desired, but the pressure obtained with these is low, and most enthusiasts prefer the cycle pump. If a built-in pump is preferred it need not be arranged inside or part of the tank, but sited in any convenient place in the hull.

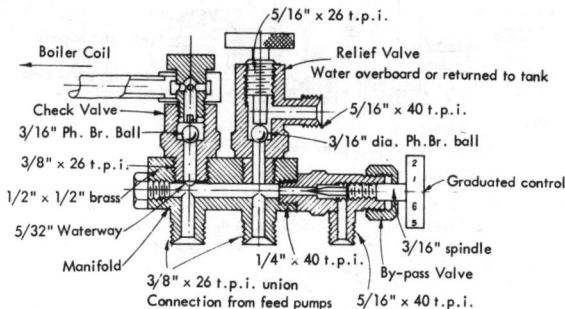

Fig. 9.11   Manifold: check relief and by-pass valves
($\frac{1}{2}$ full-size)

Air valves of the Primus or similar pattern can be incorporated in the tank filler-cap if this is made of adequate size. A simple pressure release will also be required to release air in emergency or for refilling.

A check valve in the water feed line between pump and boiler is used in most plants, also a water release and by-pass. All these oddments tend to make the plumbing a bit untidy in most flash steamers. We would suggest that the fitting shown in *Fig. 9.11* should be considered, which provides for all the above gadgets in one unit. A block of ½-inch square brass has tapped holes into which the other components and unions fit. The two unions at the bottom receive water from hand and mechanical pumps, and the end tapping is made to fit a by-pass needle valve. The union from this valve is connected with a small-bore pipe to discharge overboard. Tapped holes on the upper side of the block receive a check valve and a controlled ball valve for water release. The banjo fitting above the check valve ball is castellated to control lift yet allow water flow. It might be considered that the by-pass could be used to stop the plant, instead of having a separate valve. This is not good practice, however, as the small passages will prevent rapid pressure reduction, and also it is undesirable to vary the by-pass setting once found.

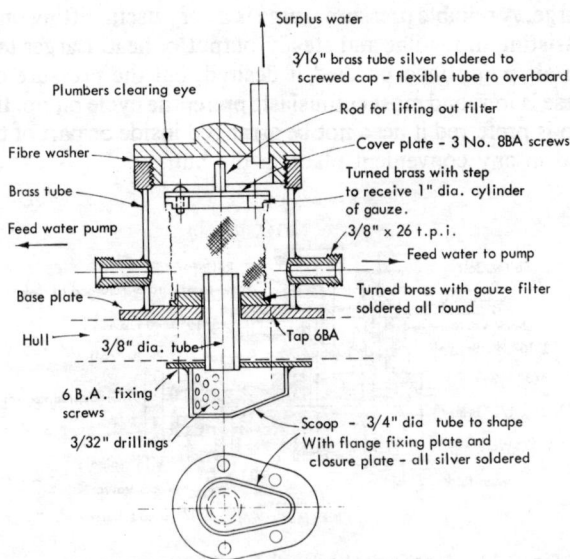

Fig. 9.12   Water scoop and filter

Model power boats have to use water which is generally far from crystal clear. In 99 per cent of regattas the pond water contains much solid matter held in suspension including vegetable fibres, grit, and the minute skeletons of pond life. Sometimes the water is salt as the ponds rely on sea-water for replenishing. The fitting of a feed water reservoir tank will enable the craft to negotiate all of these hazards without trouble. Pure rainwater can be used, and will give far less trouble with furring of pump valves and check valves.

If a water tank cannot be fitted for any reason a filter is essential, fitted between the scoop and the pumps—even minute particles may affect operation and cause undue wear.

For a fast planing boat, too large a scoop will offer considerable resistance. If it is not sited correctly it may have a steering action fatal to results in steering competitions. A central position not too far forward is usually best.

Fig. 9.12 illustrates a combined filter and scoop, the two union fittings being connected to pump intakes. Excess water and trapped air passes through the vent pipe at the top with piping exit well above the waterline. The filter is removable and the tank can be cleaned out by unscrewing three screws and the cap.

The actual size is not very critical and the design offered uses a standard brass fitting sold as a cleaning-eye fitting for sink basin traps. These are obtainable at builders' merchants. A little lathe work will remove some of the excess material before jointing to a brass tube and baseplate. The through fastening will depend on the material and construction of the hull. The scoop shown is suitable for a hard chine launch hull.

Whatever type of boat is envisaged, size and choice of propellers will vary considerably. Some ship models require a gearbox to operate twin screws of opposite 'hand', and the load provided by these would probably be greater than that of a single propeller. The actual load on the engine will affect the amount of feed water supplied by the pump, so adjustments may well be required to pump or by-pass control when experimenting with different propellers.

We would strongly urge that all those stalwarts who build and run flash-steam plants should keep a log detailing all alterations, adjustments and troubles—by such means much of the *unnecessary* experimental work may be avoided!

CHAPTER TEN

A RACING FLASH-STEAM
PLANT

The choice of a suitable design to offer the model engineer who wishes
to 'have a go' is a very difficult one. Perusal of Chapter 7 will reveal
that few engine features have become standard, also that the superior-
ity of one layout or valve gear when compared with another is
decidedly uncertain.

A decision in favour of piston valves was fairly easy, since a large
majority of successful engines have used them. But should one have
one valve or two? What about special drives? Which is best, single
cylinder or twin cylinder? In the end, we have decided to use Jim
Cruickshank's design, first described many years ago, but which is by

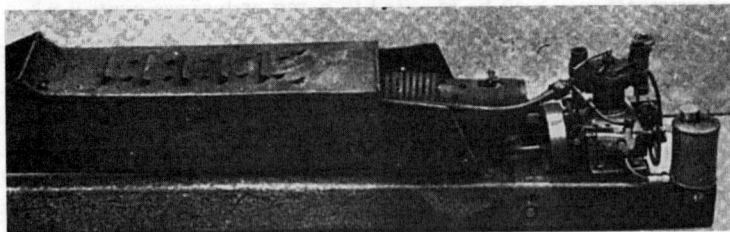

Plate 27   Jim Cruikshank's engine and boiler mounted on
a test bed

no means antique. Our reasons include the need for a design of engine
which has been tried and tested, suitable for either A-class or B-class
hydroplanes. One which is robust yet not too heavy and built without
too much difficulty. Last but not least, the engine has been made
available for examination by Jim Cruickshank, who has hopes of
installing it in a new hull in the near future.

142

Plate 27 shows the engine and original boiler mounted on a base made of steel angle for bench testing. An extra-heavy external flywheel is fitted for this purpose, and some *hundreds of hours* running have been recorded with the plant in this form. A closer view of the engine and blowlamp is shown in *Plate 28*.

Plate 28   The single-acting engine showing the arrangement

Although fairly full dimensions were given in the previous chapter, main dimensions only are quoted here in order to provide latitude for constructors. Machining methods are suggested which should enable model engineers of average experience to make the engine. In any case, it is apparent that most designs tend to be modified to some extent by their builders. Considerations include material sizes available and limitations of equipment, or simply that more experienced chaps insist on working in some ideas of their own!

Specification

Bore $\frac{27}{32}$-inch, stroke $\frac{7}{8}$-inch single-acting, separate piston valves for admission and exhaust. Split crank incorporating internal flywheels and roller-bearing big end. Crankcase of aluminium alloy split on horizontal centre line.

Referring to *Figs. 10.1* and *10.2*, it will be noticed that the crank-case has square sides and appears somewhat functional. The reason for choosing this shape was to facilitate machining and to make for easy attachment of auxiliaries such as water and oil pumps and their gearing.

Some features of design could be modified. For example, the cylinder screwed into the crankcase could be changed for the traditional

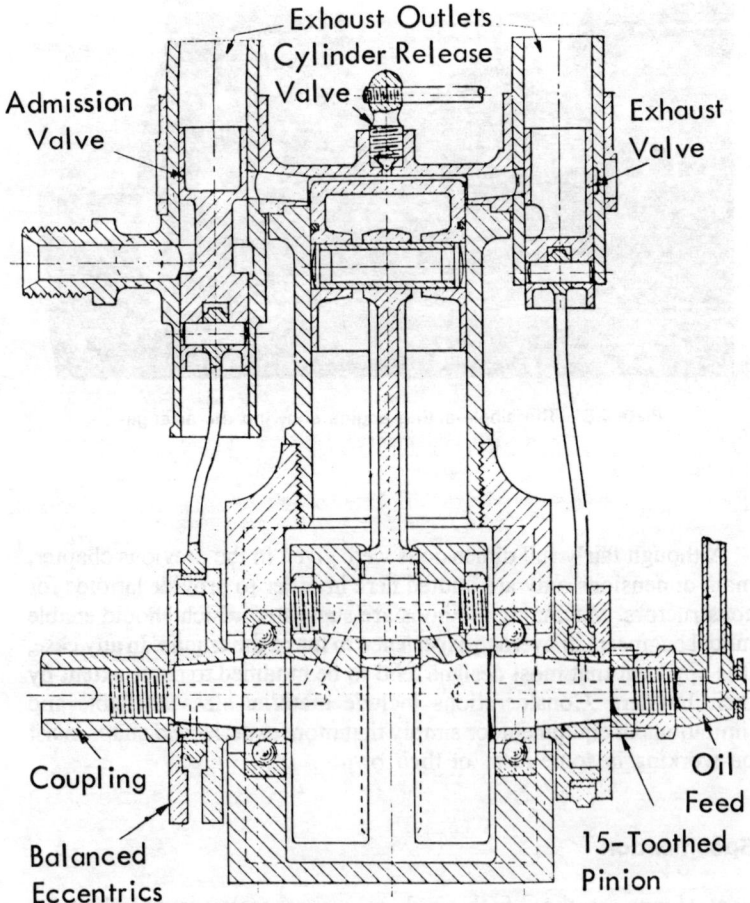

Fig. 10.1 Full-size diagram of sectional side elevation of the engine. External fittings are omitted. The oil feed is a banjo union

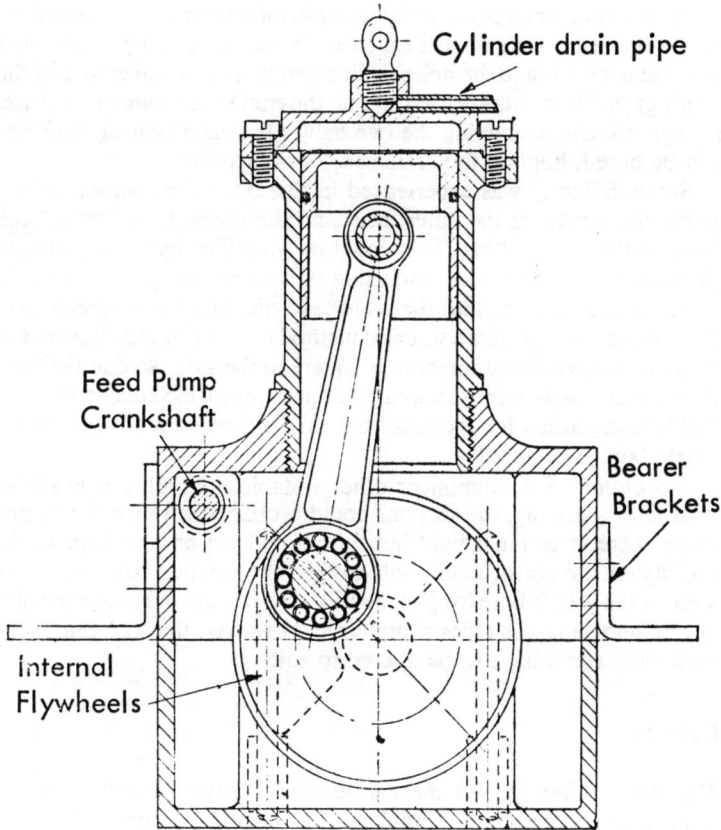

Fig. 10.2   The corresponding diagram (Fig. 10.1) of the
end elevation

flanged pattern if screw-cutting was not possible. Such modifications
will be discussed in the description of the main items.

Crankcase

The original crankcase was made from a simple casting made by
melting some scrap aluminium in the domestic boiler. An open sand-
mould can be used for such a simple shape, providing that sufficient
allowance is made for machining, or the metal may 'be poured in a
rectangle of thin sheet metal stood endwise in a tray of dry sand. The

exterior is machined oversize by turning, then sawn on the crankshaft centre-line ready to remove the interior by end milling. The sawn faces are machined flat and the holes drilled and tapped on either side of the bearings. Sunken nuts are used so that the engine can stand on its base, if required. On assembling the two halves the main bearing housings can be bored, holding the crankcase on a faceplate.

Some difficulty was experienced in this operation, because of the reduced diameter at the outer faces and the necessity of boring both holes at the same setting. This could be simplified by boring straight through to the bearing size, and using exterior retaining plates suitably spigoted and recessed into the crankcase enough to give clearance to the eccentric straps. If the screwed method of cylinder attachment is to be used, it is vital that the base is square to the sides so that the faceplate can be used for the screwcutting operation; 1-inch diameter × 24 T.P.I. is suggested for the latter but this is open to modification if a larger bore is required.

If a chunk of daralumin or other suitable light alloy is available instead of a casting, the sections could probably be reduced here and there, thereby saving weight but this should not be overdone, or the rigidity of the construction will suffer. The original ball races used were of the $\frac{3}{8}$-inch bore by $\frac{7}{8}$-inch O.D. size, but there is ample metal to accommodate larger races if preferred. However, the stud centres on each side would have to be altered to suit.

Cylinder

Materials   Cast iron or meehanite upper flange $1\frac{1}{2}$-inch diameter, lower part screwed to suit crankcase; as mentioned above, the latter could be modified to a flange bolted to the crankcase, if required, but the screwed pattern was adopted in an attempt to obtain accuracy. The final machining of the bore was carried out with the cylinder screwed into the crankcase and mounted on the lathe faceplate. An adjustable reamer with all but three blades removed was used for the final finishing, but the lapping methods, familiar to readers of *Model Engineer*, would probably give an even better result.

It is suggested that if an A-class hydroplane is envisaged, the bore might well be increased to one inch. No problems are involved in doing this, apart from modification of the exterior dimensions to suit the bore, and alteration of the offset shown on the eccentric straps.

Crankshaft and connecting rod

As designed this component may present the constructor with the most difficult problems. It is of the typical motor-cycle pattern having separate halves joined by a crankpin including tapers and sunken nuts. *It is essential that the crank throws are accurate and that the tapers fit perfectly.* The throws can be paired by final boring on a crank-turning fixture or Keats vee-angle plate, and the taper finished by a taper D-bit made at the same tool setting as the crankpin tapers so that the tapers are exactly the same.

Some constructors may boggle at the idea of making the roller-bearing big-end, as the rollers operate on the crankpin and eye of the con-rod. If preferred a Torrington B 66 needle roller can be used. For this the crankpin is increased to $\frac{3}{8}$-inch diameter from $\frac{5}{16}$-inch and the con-rod eye made smaller from $\frac{5}{8}$ inch diameter to $\frac{9}{16}$ of an inch, so that the bearing can be pressed in. Although the Torrington needles work directly on the crankpin, they are smaller than the $\frac{3}{32}$-inch diameter ones: Since the Torrington race is $\frac{3}{8}$-inch wide, the recesses in the crank webs would have to be deepened by $\frac{1}{32}$ of an inch, but this should not materially affect the holding of the tapers.

The best material for the webs and pin is high tensile steel; the original shaft was made from nickel-chrome steel obtained from a vehicle half shaft. Trial and error methods were used to find the best heat treatment, and it was found that heating to cherry red and quenching in equal parts of paraffin and engine oil gave good hardness combined with considerable toughness.

Mild steel *could* be used for the webs and journals, but some kind of hardened surface is essential for the crankpin. Case-hardened nickel steel would be suitable, or experiments could be made with H.T. steel from a suitable bolt.

If the 'Torrington' needle race is to be used, the con-rod may well be made from dural, thereby saving in weight. Otherwise, a similar steel to that already mentioned is required. The gudgeon pin is $\frac{1}{4}$-diameter drilled $\frac{3}{16}$ of an inch right through and having end pads of soft metal.

Piston

The piston is of the high tensile steel (already mentioned), having brazed-in gudgeon pin bosses and a single narrow ring, and finished to a very small clearance. Aluminium alloy pistons have been used successfully in flash-steam engines and it would be possible to use this material, but experiment would be required on the correct clearances. Pistons should be made as light as possible with skirts not thinner than ·025 of an inch for steel and ·035 of an inch for aluminium.

Head and valves

Originally the head was machined from solid mild steel with valves operating directly on the steel. In the course of development, however, part of the original steam chests were removed and cast iron liners used; liners were press-fitted as the method allows fairly easy replacement should excessive wear occur.

The steam admission to the valve caused some side thrust which resulted in severe wear on the valve; an annulus machined in the liner improved matters. Better still, a tangential entry as well could be incorporated quite easily when the brazed-on entry pipe is fitted. Inlet valve is $\frac{3}{8}$-inch diameter $\times \frac{7}{16}$ of an inch stroke. Exhaust valve $\frac{3}{8}$-inch bore $\times \frac{1}{4}$-inch stroke. Gudgeons are $\frac{1}{8}$-inch diameter with soft end pads. The admission valve also acts as an extra exhaust outlet.

An alternative to making from the solid would be a brazed up construction, and this would be suitable if a brazing strip of fairly high melting point was used. Low-melting point silver solders are best avoided because there might be occasional excessive steam temperature.

Eccentrics and straps

These make up into simple needle bearings. But since both have direct pressure-feed lubrication, a suitable plain combination such as case-hardened sheaves to dural straps would be little less efficient. Eccentrics are made in two parts, one half comprising the sheave and one side cheek, the other having a side cheek only. Balance weights are incorporated in these sheaves, and are fitted to tapers drawn up by the nut securing the pump pinion and the propeller shaft coupling, respectively. Any timing positions can thus be selected by moving the eccentric position and pinching up. Tapers: 10 degrees included angle are suitable.

Other details

Two further views of the engine, in *Figs. 10.3* and *10.4*, show the attachment of the water and oil pumps and their respective gearing. Somewhat stouter gears than those depicted would be an advantage, since a boiler capable of supplying very high pressures would probably cause some distress in this department if the effects on other plants are taken into consideration. Ratio is 5:1 (75:15), but anything near this would do. Eccentric bushes fitted direct into the crankcase enables fine mesh to be obtained without difficulty.

It is not proposed to describe the pumps since they are quite conventional and should be constructed according to the principles laid down in Chapter 7. One comment should be made concerning the

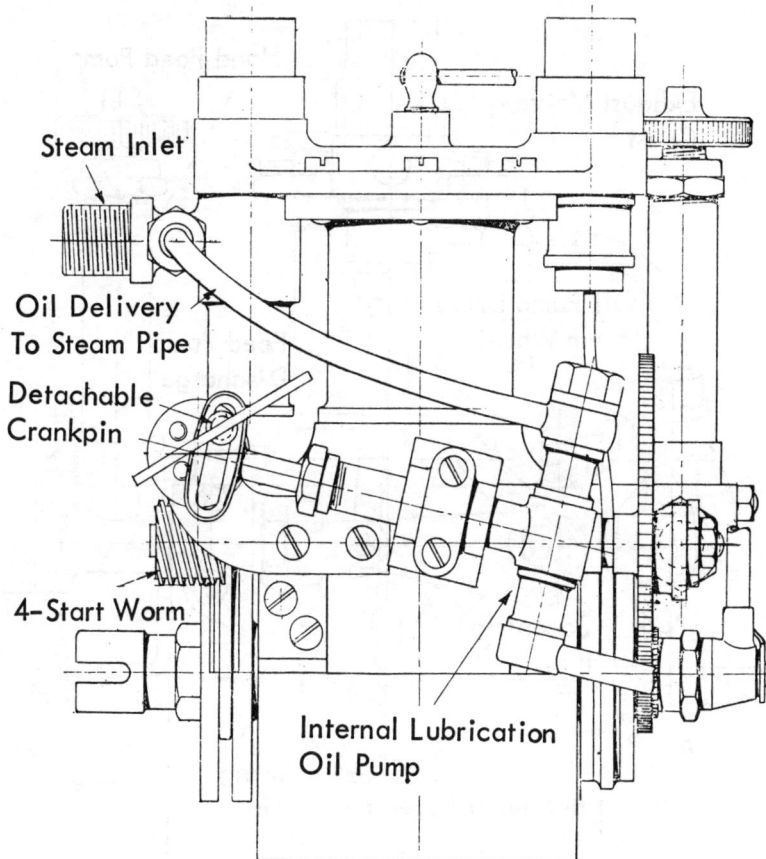

Steam Inlet

Oil Delivery
To Steam Pipe

Detachable
Crankpin

4–Start Worm

Internal Lubrication
Oil Pump

Fig. 10.3   Full-size diagram of the starboard-side elevation
of the engine

vertical hand pump—this has proved somewhat difficult to operate without an assistant firmly holding the boat. One of the alternative patterns might be preferred. In any case, a method of isolating the hand pump or of locking the plunger must be arranged. The vertical pump has a few threads at the end so that locking can be done by twisting the operating handle. Water pump attachment is by a 'plummer block', allowing minimum clearance to be arranged on the ram if stroke is changed.

Oil pump drive is by worm and wheel (Bond's) 6 : 1 from pump

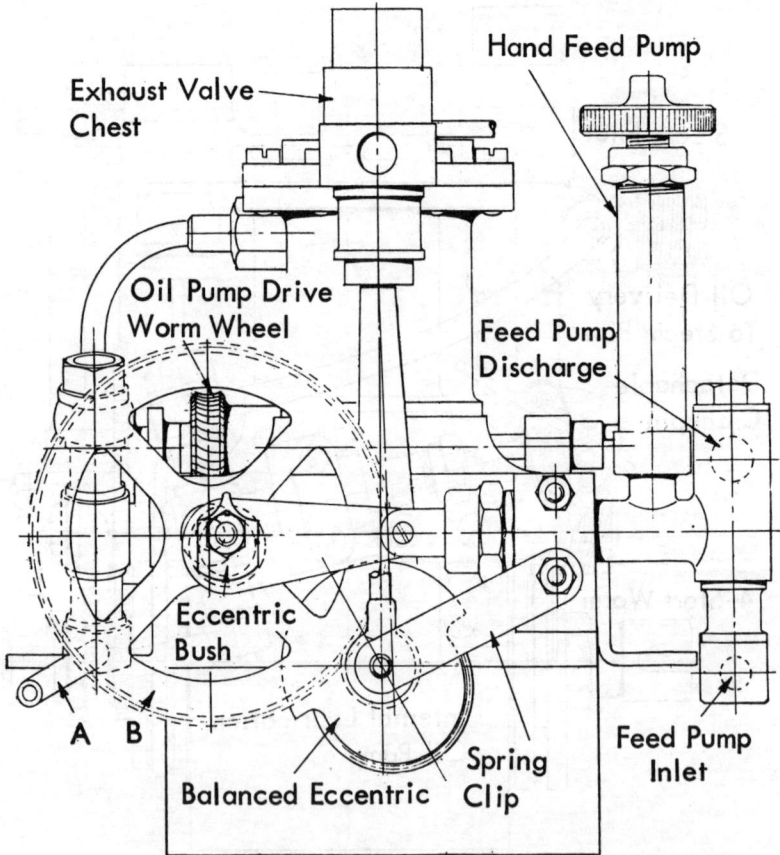

Fig. 10.4   Forward end elevation of engine. The oil pump inlet is at A, the 75-toothed spur gear at B

shaft. The detachable crankpin on the main oil pump is to enable it to be 'primed' before starting the plant; a feature found necessary because of the time lag in the oil reaching the steam chest. A valve seizure was caused by this delay on at least one occasion.

Oil is fed into the crankshaft end by means of a banjo held on by a small spring clip. It should be noted that it is vital to line up the oilways of the crankpin on assembly so that oil reaches the big end and far side eccentric. See plan view, *Fig. 10.5.*

Boiler and firing

There is little doubt that the Bamford venturi-type boiler would produce the most steam and therefore the most power, but it really

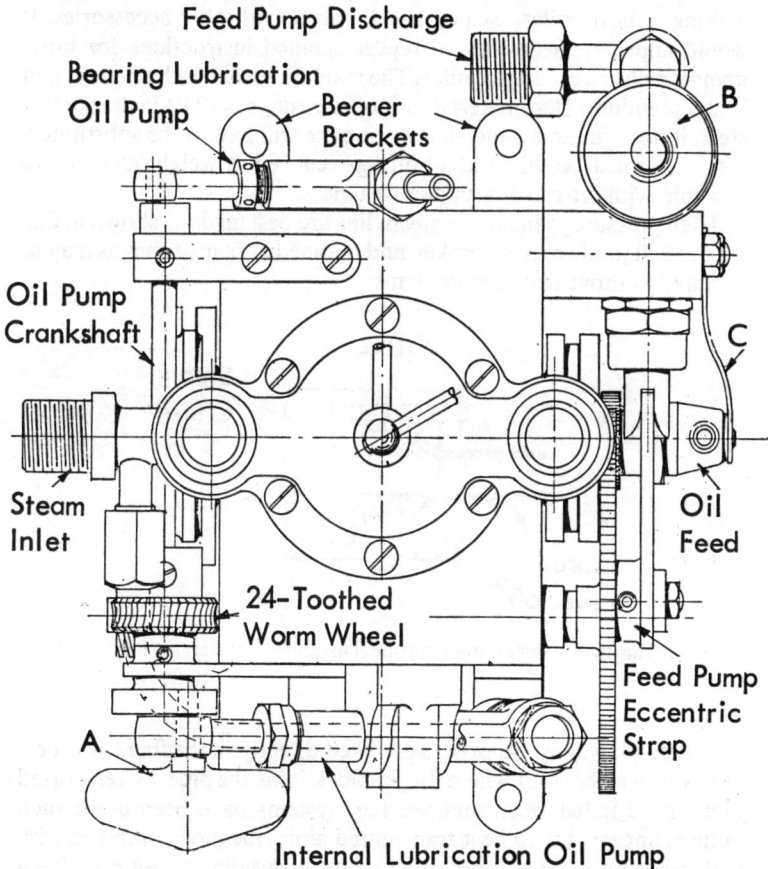

Fig. 10.5   The plan of the engine. The oil feed is by a banjo union held in place by a spring clip. A indicates the position of the detachable crankpin, and B the position of the hand pump

requires a pumped fuel supply to work well. The constructor who is willing to tackle this could easily incorporate a $\frac{3}{16}$-inch bore pump alongside the main water pump, connecting the two rams by a bar so that they operate together. Petrol/paraffin mixtures are mostly used for fuel—the petrol content making it easier to warm up the lamp.

The Bamford venturi boiler has already been referred to in Chapter 5 without quoting specific dimensions. There appears to be some latitude in the precise shape since Frank Jutton's version is far from an exact copy.

Very full constructional details are given in the previous chapter for making a flash boiler, casing, check valves and other accessories. It would appear unnecessary to repeat detailed instructions for those items which are broadly similar. The main difference is that low weight is at a premium! Best material for boiler casings is ·010-inch stainless steel, if not available, mild steel or tinplate will have to be substituted. Asbestos must be thin and all fittings reduced in weight as much as possible without causing undue weakness.

High pressure joints in the steam line are best made as shown in *Fig. 10.6*. Such joints may be broken and rejoined as many times as may be required without leakage problems.

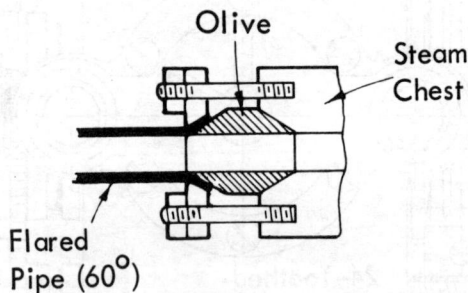

Fig. 10.6   Recommended union for joints in the steam line

Where blowlamp vaporisers are sited directly in the flame it is best to have a welded joint where the jet block joins the pipe. Silver-brazed joints have failed when pumped fuel systems have been used; such failures appear due to heat transmitted along the pipe combined with high pressure created by the fuel pump. If welding is not possible a screwed and brazed joint would be suitable providing the tube wall is thick enough to take one of the fine *M.E.* threads.

The vaporising tube should be of stainless steel or thick-walled mild steel, but not copper or brass. The last metals may only be used where the coil is indirectly heated. All tube used for vaporisers must be seamless for obvious reasons.

Ideally, all casings should be lined throughout with thin asbestos but this may not be possible if weight limits are getting a bit tight. Asbestos can weigh quite a lot! If thin asbestos cloth is obtainable instead of millboard, it is first choice since the millboard tends to flake after a while.

Small self-tapping screws are excellent for making joints in thin sheet metal, and a scriber point pushed through the thin metal will be sufficient to start the screw besides offering a larger area for the thread.

The flue exit should not be made too large as it is desirable to retain as much heat as possible yet without baffling the gases. It should be fairly easy to arrange an adjustable flap to control the exit, if thought desirable.

Coiling of boiler tubing is fairly easy if a single-coil boiler is chosen. Double concentric coils as in the venturi boiler are more difficult, but as joints will probably be required somewhere it may be possible to wind separately and make the joint at one end.

A word here about joints in boiler tubing—welding is the first choice, but welding equipment and 'know how' is a requirement for welding joints in thin tube. Brazed joints with a sleeve last a fair time, especially if the join is sited away from direct heat of the flame.

As an alternative to the above boiler, we suggest a variation of the 'Bamford Mark II' boiler. A prototype has been made which appears to work well, and it may be fired with a pressurised lamp having an external vaporiser. The principle is that of a flame blowing into a closed casing—all air required for combustion is taken through the lamp flame. It is based on the boiler used in Jim Bamford's *Bebug*, but having a single large burner instead of three. The example depicted has two concentric coils of tubing, and if a single coil is preferred the length overall will be somewhat greater. Casing is circular as in the venturi boiler but other shapes could be tried and would no doubt prove just as successful.

If a 'burner aft' position is required the uptake will require modification so that the exit is not subject to draught caused by the motion. Refer to *Fig. 12.7*, which illustrates the Mark II Bamford boiler.

A single burner has been chosen partly for reasons of simplicity, and also to reduce the diameter of the casing so that a modern planing hull without excessive beam could be used. Space is also somewhat limited in a B-class hydroplane and any measure that helps to reduce weight and bulk is welcome, *Fig. 10.7*.

Boiler tubing should be $\frac{1}{4}$-inch O.D., wall thickness 20 or 22 S.W.G. If the exchequer will not run to stainless steel, mild steel will have to do. If weight problems allow, the thick-walled tube has a considerably longer life.

For a single-coil boiler the tubing is wrapped around a former about 2½-inch diameter. The easiest way would be to have one coil without a cross coil or any loops placed in the centre. This, however results in a rather long boiler if 30 feet of tube is to be used. It is desirable to space each loop at least ⅛ of an inch apart, and this adds to the problem.

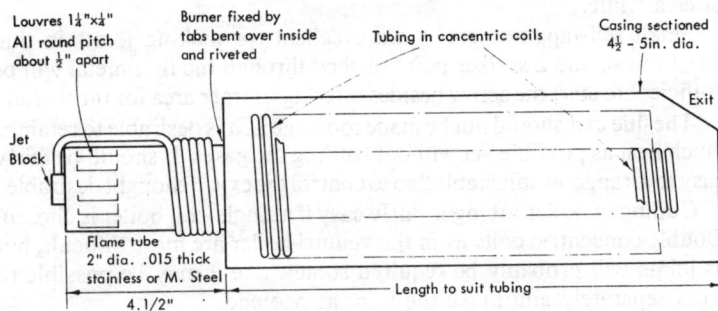

Fig. 10.7   Enclosed flash boiler with single burner

We suggest 20–25 feet of tube for B-class and up to 30 feet for A-class. It will be understood that the exact length of the lamp and boiler will depend on how much tubing is used, and its arrangement. The concentric coil boiler is much shorter but larger in diameter.

As explained in the previous chapter, a check valve and a water release fitting will be required at the feed end, and a union at the steam exit to connect to the engine.

The blowlamp is of conventional style with one important exception—holes or slots are not used for admission of air, instead adjustable louvres are cut in the end of the flametube so the amount of air is easily controlled by bending the flaps. A fixed jet of ·032-diameter can be tried initially, or an adjustable jet made up—in this case the jet should be ·040-inch diameter thus allowing a fair range of adjustment.

Since the vaporising tube is only heated indirectly, it is not vital to use stainless steel. Mild steel, copper or brass will be equally suitable providing that it is seamless. The reason for chosing an external vaporiser was to make it possible to use petrol or blowlamp fuel. However, if it is desired to use paraffin the tubing will probably require placing in the direct heat of the flame within the boiler casing. Brass or copper tube is then unsuitable.

If the flame fails to stabilise in the burner tube, a rod of mild steel about ⅛-inch diameter sited about halfway along will do the trick. It is easily replaced when burnt out.

Pressurised fuel containers should be of silver-brazed construction throughout, and of light gauge seamless brass or copper tube. Rolled up containers of thin mild steel sheet should have a clenched joint along the seam, flanged ends, and all ferrules used are best made with a small lip so that they can be inserted from the inside and brazed before the ends are put in position. Containers should be given an hydraulic test to twice working pressure.

Some latitude is possible as to container dimensions according to the available space. However, the largest size is to be preferred, so that only a small amount of fuel is carried yet a large volume of compressed air.

One of the best air valves is of the Primus pattern, which can be 'built in' to a filler cap. Such valves used to be readily obtainable, but since the use of conventional blowlamps has declined in the building trades, stocks of spares are not very high in most ironmongers. The old-fashioned bicycle valve using a rubber sleeve was unsuitable for blowlamp work as fuel attacked the rubber, causing it to swell. Modern bicycle valves seem unaffected by petrol, and could be used instead if the Primus type cannot be obtained.

We have mentioned elsewhere that many modern boats carry water instead of scooping it from the pond. A light tank can be made from ·006-inch brass 'shim' soft soldered together, and of any convenient shape. In the matter of weight the position is not as bad as might appear—scoops, overflows, reserve tanks and filters are all eliminated and clean water is used by the pumps instead of the semi-solid matter that passes for water in most ponds! Class weight limits are quoted as with empty fuel and water tanks so there is no problem here.

A tank to hold 1 lb. water for B-class, and about $1\frac{1}{2}$ lbs. for A-class, will be ample for a long run. Tanks should be sited on or near the centre of gravity so that the weight distribution is not materially affected as the water is consumed.

CHAPTER ELEVEN

SOME EARLY FLASH STEAMERS

There are some honoured names in the very early history of flash-steam hydroplanes. The achievements of these enthusiasts were remarkable when it is remembered that they were real pioneers and literally had to find out everything for themselves. Although it must be admitted that today some aspects of flash plants provoke controversy, the work of these men still remains in the form of basic ideas and plant layouts which are virtually unaltered over 60 years later.

There has been no lack of criticism, but in spite of sound theoretical reasoning no one has appeared with a practical demonstration to prove their arguments. A good example may be recalled from Chapter 5, relating to firing methods: 'torch-type lamp versus diffused-flame burners'. The torch pattern—even if in modified form—still reigns supreme!

We would like to pay tribute here to all of those pioneers—some of whom never took part in regattas—and others, who carried out their experiments as lone hands without public recognition of their work. It is not possible to list all who deserve mention, but it was thought that a few brief notes on some early craft and their constructors would not come amiss. The dates given are approximate only.

Herbert Teague and V. W. Delves-Broughton, were an artist and mechanical engineer, respectively, and to them falls the honour of having built the first flash-steam hydroplane on record, *Folly* (1908). This craft had a twin-cylinder, single-acting engine, overhead slide valve operated by bell-crank and rear-facing blowlamp firing a flash boiler sited in a rather high casing. The hull had two steps and the best speed approached 20 *mph*. A later craft *Incubus* was a little faster, but no other hydroplanes were built at that time probably due to the outbreak of World War I.

G. D. Noble of Bristol was also a very early exponent, together with his brother S. S. Noble. Several different craft were built within a few years but the best known flash steamer was the original *Bullrush* (1912). A contemporary photograph shows a remarkably modern style, using a boiler fired from aft with an unusual exit for the flue which discharged the spent gases aft. The engine was another slide-valve job, twin-cyclinder S.A., $\frac{15}{16}$-inch bore × 1-inch stroke not unlike many more modern engines of this type. A later craft described in 1923 had an engine of the vee-four type—the only known example applied to a flash steamer. In fact, there is only one other known 4-cylinder engine of any sort and that was a flat four by Mr. Willis of Dublin. The 'Noble' vee-four was fully described in December 1923, with detailed drawings, but it is not thought that any further examples were constructed. It was a most interesting design and with minor modifications would serve as the basis for a powerful engine for a 'Class A' steamer.

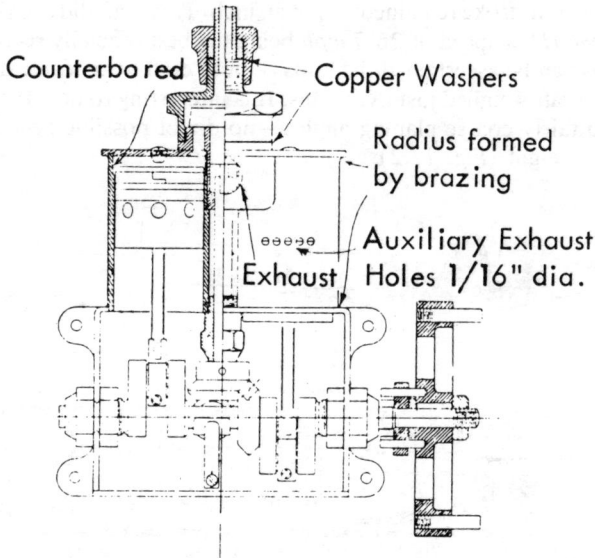

Fig. 11.1   The engine used in *Irene III* by H. H. Groves

A breakaway from the overhead slide or piston valve was made by H. H. Groves, who was also an experimenter with flash-steam model aircraft. This exponent was a master of lightweight design and construction, thus the metre long hydroplane *Irene III* (1913) weighed only about 7 lbs. The chief item of interest was the engine. This was a departure from the vertical twin arrangement, as the cylinders were arranged horizontally and brazed directly to a light crankcase. Instead of a slide or piston valve a very simple disc rotary valve was situated between the cylinders, exhaust was by uniflow ports and a hollow drive spindle of the valve. This valve arrangement presented some difficulty regarding the drive as bevel gearing had to be fitted between the crank throws and a half bearing used to prevent deflection (*Fig. 11.1*). The boiler had 23 feet of $\frac{3}{16}$-inch steel tube in a single coil, enclosed in a perforated cylindrical casing. Firing was by single burner $1\frac{1}{4}$-inch diameter by 4 inches long.

Mr. Groves was one of the first experimenters to realise the combustion problems of boilers. Reference has already been made to the boiler casings he adopted.

Another great exponent of flash steam was F. Westmoreland. His first steamer was a replica of *Irene III*. This new craft was *Evil Spirit*. Although almost identical, the engine bore was increased to $\frac{12}{16}$ from $\frac{11}{16}$ of an inch and stroke remained at $\frac{5}{8}$ of an inch. This craft did even better than *Irene III*, a speed of 26·7 *mph* being the best officially recorded. As far as can be ascertained the hulls of both craft were identical, but the later craft weighed just over 8 lbs. It is interesting to note the deep step and fairly coarse planing angles—no doubt possible because of the light weight (*Fig. 11.2*).

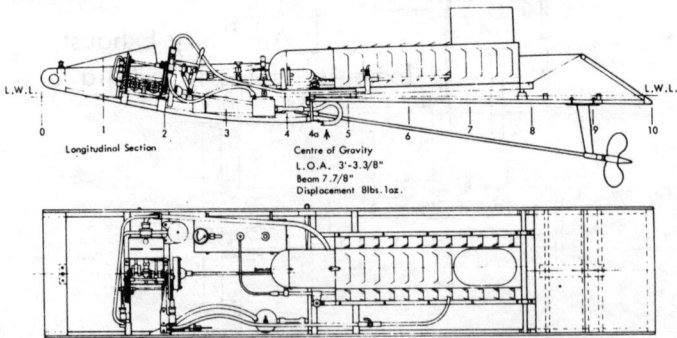

Fig. 11.2   *Evil Spirit* by F. Westmoreland

A later effort by F. Westmoreland was the hydroplane *Mystery*. The engine represented a completely new line of attack, since it was a twin-cylinder poppet valve job designed by Thos. Hindle(an expert on flash steam and steam cars. The engine was of the uniflow type, and although no records were broken, it put up some very fast runs in practice. Mr. Hindle informed us recently that the plant operated at very high pressures, and if the engine stopped on the course with the lamp still going, the boiler invariably burst. This boat ended its career by hitting a full-size rowing boat and turning it right over!

The hull design of *Irene III* had a strong influence on several later hydroplanes. A series of boats by A. Norman Thompson bearing the name of *Sunny Jim* (1922) were very similar in appearance. The engine used in the most sucessful craft was a piston valve job, with overhead piston valve and bell crank and eccentric operation. The boiler was very similar to the 'Groves' pattern having inner and outer casings. Over 30 *mph* was achieved many times. On more than one occasion the pull of the boat caused either hook or line failure resulting in considerable damage.

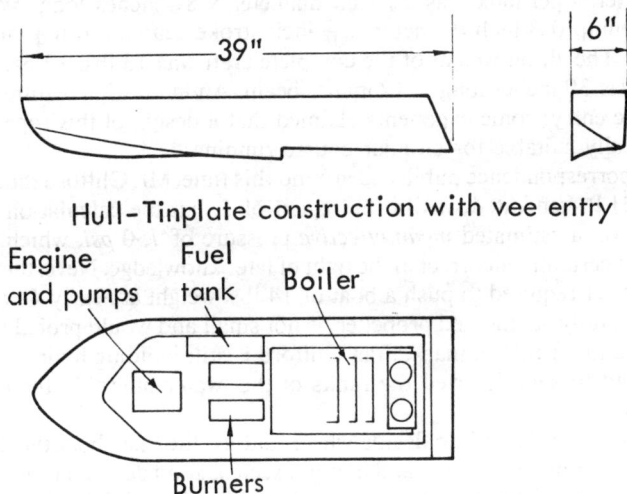

Hull-Tinplate construction with vee entry

Fig. 11.3   Layout of *Chatterbox III* by S. Clifford

A new engine having semi-rotary valve, described in 1926, appeared to be less successful as no record can be found as to its subsequent performance.

Now we come to the redoubtable *Chatterbox III* (1925) by S. H. Clifford. This was the last of several boats bearing this name—the first one not being very successful. Number two in the series achieved 36·4 *mph* in the *Model Engineer* speed boat competition for 1924. On building a new hull, which became *Chatterbox III*, and installing the same plant, 37 *mph* was attained in 1925 41·85 *mph* in 1926, and finally 43·84 *mph*. This was claimed as a world record and remained unbeaten for nearly ten years by any boat, petrol or steam (*Fig. 11.3*). The speed was all the more remarkable when it is considered that even the fastest petrol-engined craft were still hovering in the mid-twenties at this time. Indeed, there were no entries of petrol-engined boats in the *M.E.* competition several years in succession.

The increase of power when compared with contemporary flash steamers would appear to relate to the successful boiler, since the engine was another of the type that was almost standard at this and indeed of later periods. Details included $\frac{7}{8} \times \frac{7}{8}$-inch twin S.A. overhead piston valve, bell-crank operation.

No less than 40 feet of steel tubing—30 feet of it $\frac{5}{16}$-inch diameter—was arranged in two parallel coils and fired by two blowlamps $1\frac{7}{8}$-inch diameter. Fuel tank was $2\frac{3}{8}$-inch diameter $\times$ $8\frac{1}{2}$ inches long. Water feed pump $0 \cdot 3$-inch diameter $\times \frac{7}{16}$ inch stroke geared down $\frac{1}{4}$ engine speed. The all-on weight of the complete craft was 13 lbs. 12 oz. The hull was 39 inches long $\times$ 12 inches beam. An interesting feature was the vee entry; some exponents claimed that a design of this type was not really suitable for circular course running.

In correspondence published around this time, Mr. Clifford thought the I.H.P. generated was in the region of 5, but the calculation was based on a estimated *mean effective* pressure of 750 *psi*, which was almost certainly incorrect in the light of later knowledge. Nevertheless, the power required to push a boat of 14 lbs. weight at nearly 44 *mph*, using a fully submerged propeller, is not small and would probably be around the 2 B.H.P. mark. Stan Clifford is still building hydroplanes but, sad to say, deserted the ranks of the 'water otters' in the early 'thirties.

It is rather difficult to decide where to draw the line about the dates of 'Early Flash Steamers', and it might seem a good demarcation point to end this chapter with those built before World War II. Some constructors, however, commenced activities in the 1930s and continued well into the post-war period, and thus are due to be mentioned in the next chapter.

One exponent who did much to keep flash steam interest alive in the 'thirties was A. Martin, of the Southampton M.P.B.C., with his *Tornado* series of hydroplanes. Mr. Martin was of the school of H. H. Groves; his ability in lightweight construction with high output was truly remarkable.

All the engines of the series were single-cylinder, single-acting jobs, which made a change from the vertical twins that had tended to be an almost automatic choice for many years.

The first successful boat was *Tornado II* (1933), which is thought to be one of the lightest C-class steamers on record. All on weight was only $4\frac{1}{2}$ lbs. The plant was very simple and neat, having an engine with

$\frac{5}{8}$-inch bore and stroke with eccentric-driven vertical piston valve of only $\frac{3}{16}$-inch diameter. Although an oil pump was fitted for cylinder lubrication, no bearing pump was used, external lubrication consisted of "going round with an oil-can". The boat was unstable at anything over 26 *mph*, and *Tornado III* (1935) was then built. This hull was modelled on the lines of Mr. Rankine's *Oigh Alba*, and the new hull soon showed improved speeds and better stability. A speed of 30 *mph* was exceeded during several runs. The last boiler was 11 feet of $\frac{1}{4}$-inch diameter tubing plus 8 feet of $\frac{5}{32}$-inch diameter, and by now the engine was lubricated by main bearing and big-end oil drips.

Tornado IV (1937) was a Class C (now Class B) having a larger engine and twin burner lamps. The general engine design was similar to the original, but with $\frac{7}{8}$-inch bore and stroke with full bearing lubrication by means of an ingenious dual-purpose oil pump. 'Home-made'

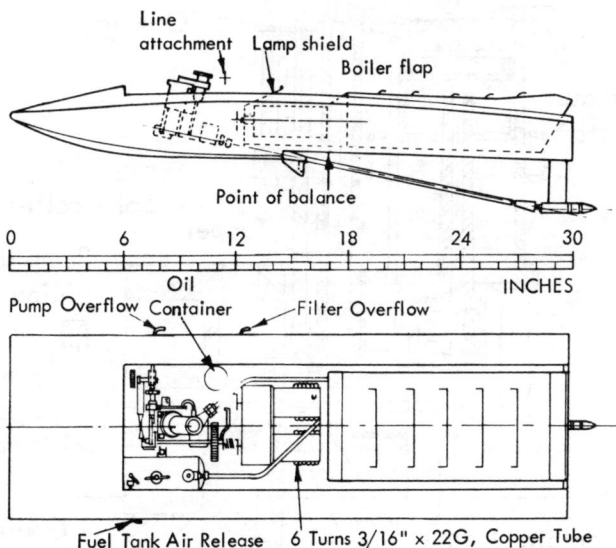

Fig. 11.4   Elevation and plan of *Tornado IV*

roller bearings were fitted to eccentric, big end and main bearing. A much larger boiler was employed, having 26 feet of $\frac{1}{4}$-inch diameter tubing and twin lamps with burners $\frac{3}{8}$-inch diameter. 'Combustion flaps' were a feature of the boiler casing (*Fig. 11.4*).

This craft was the fastest of the series, 40 *mph* being exceeded on many occasions—unfortunately, none of the performances which exceeded 40 were at regattas, the best regatta run being around the 30 mark. The engine revs peaked at 11,000. Calculation was possible by counting the strokes of the oil pump against a stop watch—every stroke of the pump produced a blue haze in the exhaust quite easily visible from the bank.

During 1937 Mr. Martin made yet another engine to replace the little piston valve engine in *Tornado III*. This was a poppet valve uniflow engine having a small piston valve to prevent compression on the idle stroke (*Fig. 11.5*). certain details such as the crankshaft were

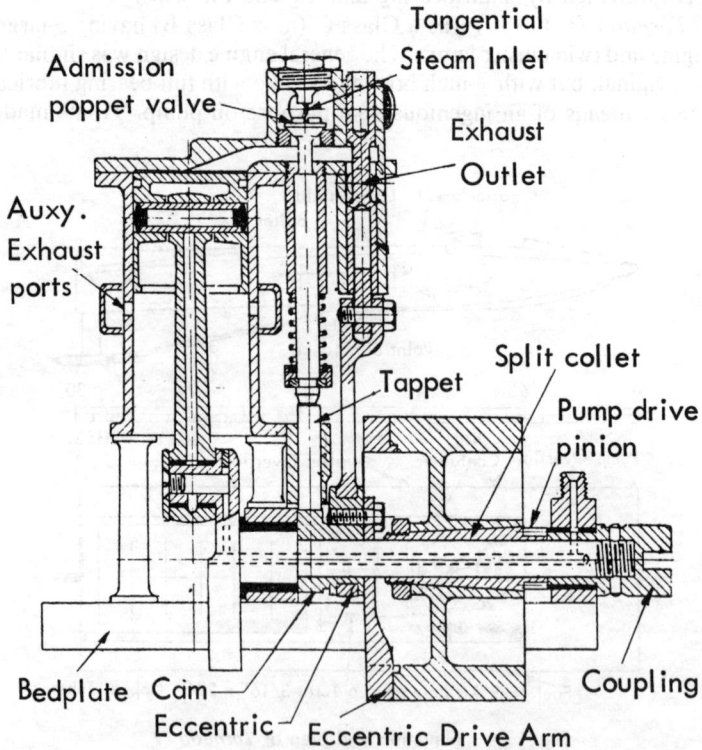

Fig. 11.5   Sectional view of the experimental poppet-valve engine

made stronger and force-feed lubrication provided to mains, big and little end. All-on weight had now risen to $5\frac{1}{4}$ lbs., but the new engine was far more efficient than the old one, resulting in a run of 30·13 *mph* for the 1937 *M.E.* Speed Boat Competition.

Mr. Martin made a successful post-war appearance from 1947 to 1950, but has not been seen around the regattas for many years.

The exponents mentioned above only represent a few of the experimenters during the period 1908–1939. For example, in the London area alone at least a dozen hopeful enthusiasts (Benson and Rayman among them!) were operational in the nineteen-thirties. Although their results were perhaps not as spectacular as some of the crack boats, they nevertheless provided much enjoyment and instruction on flash steam matters.

CHAPTER TWELVE

SOME MODERN FLASH STEAMERS

Sunbeam

The first steamer to be discussed is not really 'modern' as it was operational in the late 'thirties and early post-war period. It has, however, a good claim to be mentioned, since it achieved an offically observed 49·29 *mph*—claimed as a world record at the time. This was done by using a normal planing type of hull with fully submerged propeller.

It was built by a member of the Sydney S.M.E., Australia, so that it is the only overseas boat to be mentioned in these brief notes concerning technical details of flash steam hydroplanes.

The builder of this remarkable craft was the late Ron Cowan, and we are indebted to his colleague W. Robson for the cross-sectional diagram of the engine and other details of the plant (*Figs. 12.1* and *12.2*).

It will be noted that engine layout is straightforward, with 1-inch bore and stroke, S.A. uniflow, piston valve driven by vertical shaft bevel geared to crankshaft, enclosed crankcase of aluminium alloy. Cylinders of steel tube were welded to baseplate and valve chest. It included water-cooled valve chest and cylinders. The latter was not an original feature but a later addition, and one which effected an improvement in reliability.

The water-cooling arrangement is one which is perhaps not in accordance with steam principles, but gave good practical results. If the steam is so hot that it causes cylinder heads to glow dull red—this has been quoted as true on both model and full-size engines—lubrication is not likely to be effective. However, it would seem that the water-cooling system allowed the valve to survive. A better way would, of course, be some kind of de-superheating which did not waste heat energy—probably difficult to do in a model hydroplanes.

Valve timing of the engine is not known but the valve had $\frac{1}{2}$-inch travel, which should have enabled a fairly short cut off to be used without undue port restriction.

The boiler was 48 feet in length and consisted mostly of $\frac{5}{16}$-inch diameter tube arranged in three lengthwise coils, plus one cross coil. It

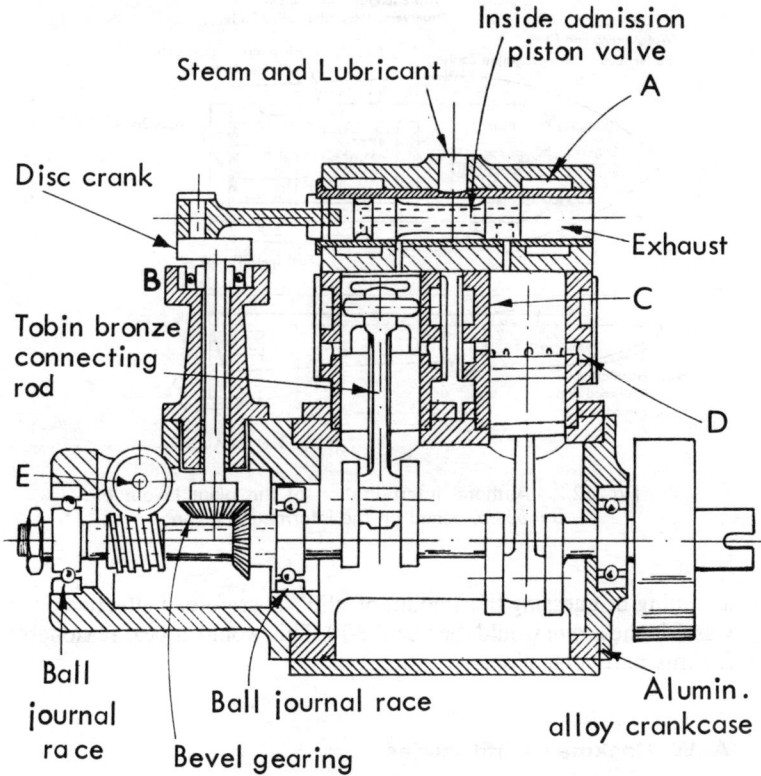

Fig. 12.1 Authors' interpretation of the engine fitted to *Sunbeam*, based on the information available. The aluminium water-cooled head and two annular chambers connected at the side are indicated by A, ball journal race at B, steel cylinders with water-cooling jackets at C, 'uniflow' auxiliary exhaust manifold at D, and worm and wheel reduction gear for pump drive at E

was fired by triple burner lamps which were sited aft and had separate air and petrol containers. These twin tanks were a unique feature, and obviously intended to minimise the pressure-drop which has always plagued this type of fuel feed. Another interesting feature was the huge propeller used—$4\frac{1}{8}$-inch diameter × 10-inch pitch—this must be one of the largest props ever used fully submerged on a boat of this type. Flash steamers usually like big props; also, a 1-inch × 1 inch twin is around 25 *cc* capacity and would develop a lot of power. 50 *mph* is equal to 4,400 feet per minute, and engine *rpm* may be estimated by

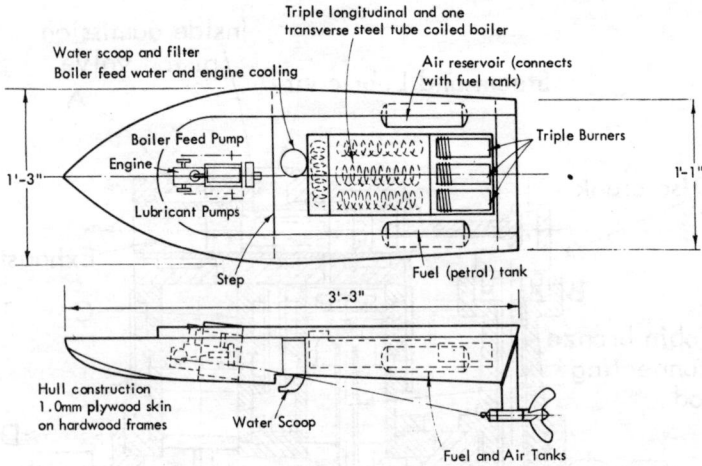

Fig. 12.2   Authors' interpretation of the plant layout of *Sunbeam*, based on the information given

assuming or guessing the amount of 'slip'. Therefore, if effective pitch was 6 inches, *rpm* would then be 8,800. This would appear reasonable for this particular plant.

A. W. Cockman's 'Ifit' series

A. W. Cockman started his flash steam experiments in the early 'thirties. After several years hard work using various rebuilt 'Stuart' engines, he designed and constructed the engine shown in *Plate 29* and *Fig. 12.3*. This engine had separate piston valves for inlet and exhaust and originally was fitted with a nitrided valve chest and valves of special heat-resisting steel. Sad to say, these matererials did not operate well and were speedily abandoned in favour of cast iron, after which no trouble was encountered.

   The engine was constructed with great care and fine workmanship. The cylinder block is brazed up steel with press-fit, cast-iron liners. Webs support the overhang of the top plate assembly. The pistons are of RR 53 aluminium alloy with gudgeons of $\frac{7}{32}$-inch diameter, and one cast iron ring, $\frac{1}{32}$-inches wide, was fitted. Clearance is ·005 inch in the bore to allow for expansion. Exceptional trouble was taken concerning lubrication. Besides the pump lubricating valves and cylinders, a separate pump feeds oil via a distributor to main bearings, big ends and all parts of the valve gear and eccentrics.

Plate 29   Twin-cylinder engine by A. W. Cockman

Ifit numbers *4*, *5* and *6* used this engine, although a large double-burner boiler replaces the original single one used in *Ifit 4*.

The boiler department was subject to experiment and Mr. Cockman reports that at one period, more speed increases were achieved by experimenting with boiler casings than any other means. The length of tubing used was only 30 feet of $\frac{1}{4}$-inch diameter, arranged in two lengthwise and two cross-coils. The casing arrangement has already been shown in a diagram in Chapter 5. It should be noted that bending the flap downwards to prevent cold air from entering the boiler directly, was an experimental success. Blowlamp burners had Primus jets opened out to ·029 inch, and about 5 ozs. of petrol were used in one run of five or six laps.

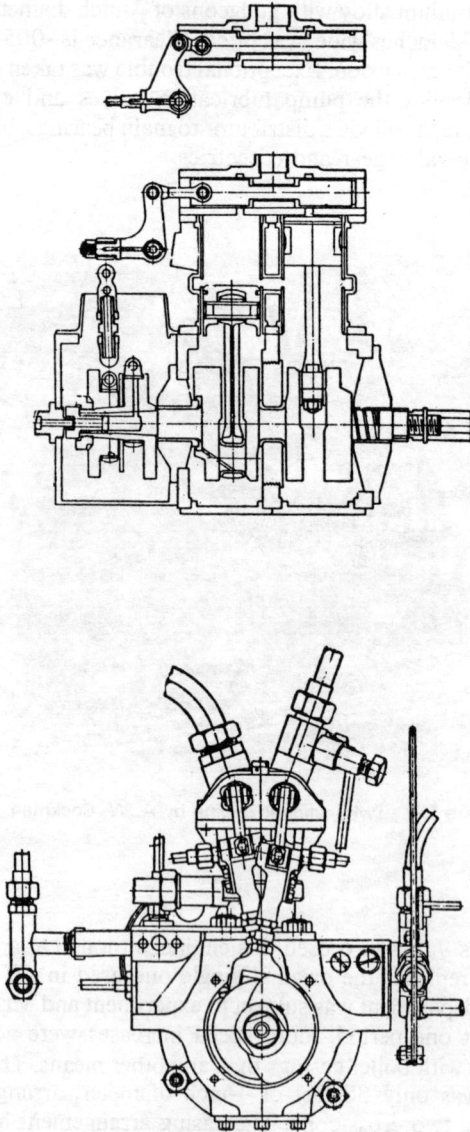

Fig. 12.3   Twin-cylinder engine with separate admission
and exhaust valves (A. W. Cockman). Top, elevation showing inlet valve with exhaust valve above. Bottom, front view
of engine with front casing removed to show valve gear

Plates 30 and 31 Two views of *Ifit 9*

Number 5 had a career lasting only about a year as the boat was unstable at speed. This led to the construction of *Ifit 6* (1937), which became one of the few hydroplanes to uphold the prestige of steam in the lean times during the late 'thirties and early post-war periods. Best speed with *Ifit 5* was 35 *mph* (over 5 laps), but *Ifit 6* managed to beat the 12-year-old record for flash-steam craft set up by *Chatterbox*. The recorded speed was 43·56 *mph*, set up at the 1938 Wicksteed regatta. Best speed in 1939 was 45·4 *mph* in the M.P.B.A. Speed Championship race.

In the early post-war period a new hull was built of the 'prop-riding' type. The plant remained substantially the same as before but with a different boiler casing. The new hull gave a considerable increase of speed and 5 laps were frequently covered in the higher fifties. At the International regatta, held at Derby in 1950, *Ifit 7* was in line to win

Plate 32   Rotary valve engine by A. W. Cockman

the 'Ford Trophy'—an all-comers race—but a pipe failure on the last of the timed laps prevented a well-deserved win. On this occasion, the speed was nearly 60 *mph*.

In 1951 a new lightweight steamer was built to conform with the 5-lbs. limit in Class C. The new boat was *Ifit 8*, and this craft was run for several years without ever reaching exceptionally high speeds, although over 40 *mph* was attained. This led to the introduction of *Ifit 9*, which used the same engine but a new boiler capable of supplying more steam. The new boat was in Class B, which has an 8 lbs. weight limit, thus allowing the larger boiler (*Plates 30* and *31*).

The same engine was used in both the latter craft and is of the rotary valve type (*Plate 32*). This is a remarkable job having a conical rotary valve partly pressure balanced and geared down 3 : 1 from crankshaft via skew gears. Although some very good performances have been seen, the gears have given some trouble. At the time of writing the boat is non-operational due to gear failure, despite the last pair being made from special steel, carefully heat-treated.

Some difficulty has also been experienced in pumping fuel as mentioned in Chapter 5, and more recently a return was made to a pressurised fuel container. The boiler used in *Ifit 9* is illustrated in *Plate 12*, and a partly finished new engine for a Class A hydroplane in *Plate 15*. This engine is, of course, the poppet valve job with desmodromic operation already referred to in Chapter 6.

Vesta II and T.N.T.

Frank Jutton' flash steam work dates back to pre-war days but the first signs of success were apparent in 1948 with *Vesta II*. The first engine used in the original *Vesta* was a horizontally opposed flat twin $\frac{5}{8}$-inch bore × $\frac{3}{4}$-inch stroke with one cylindrical rotary valve feeding both cylinders. In spite of the somewhat long steam passages the

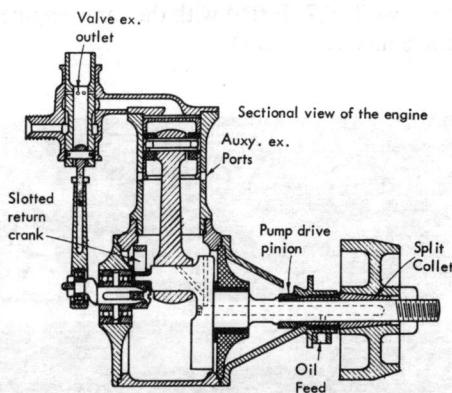

Fig. 12.4   A sectional view of the single-cylinder engine by
F. Jutton, with 'quick return' valve gear

performance was quite good especially for a conventional planing hull. This engine suffered much from mechanical trouble and a new engine was built employing the 'quick return' motion developed jointly with Bernard Pilliner, then a colleague in the Guildford club. The return crank for the valve is offset $\frac{5}{32}$ inch, apart from this feature the engine is of straightforward design, $\frac{7}{8}$-inch bore and stroke, valve chest of nickel-steel brazed to the head and piston valve of cast-iron. Cylinder is also of cast iron and piston of unhardened silver steel. A hefty crankcase is employed allowing the cylinder to be screwed in without studs. All alloy parts were machined from solid dural including the support for the outer main bearing (*Fig. 12.4*).

Vesta II enjoyed a long series of regatta placings holding the Class B record at over 50 *mph* for a number of years, but suffered from air lift which caused numerous 'flips' at anything over 45 *mph* (*Plate 33*).

Plate 33   *Vesta II* by F. Jutton

After a lay-off lasting some years Frank Jutton has made a welcome return with a new boat *T.N.T.* Fitted with the same engine, but with a 'Bamford' pattern boiler (*Plate 34*).

Plate 34   *T.N.T.* by F. Jutton

The new layout is quite unlike those of the earlier boats. Apart from siting the engine aft and the entry to the boiler venturi immediately forward, no universal joint is used, as the propeller is held about 6 inches behind the transom on a bracket made of tubes welded together. Most hydroplanes refuse to operate on straight through drives but this craft is different. Speeds of over 60 *mph* are frequent, and the experiment of carrying water, instead of scooping it from the pond, has resulted in steadier plant performance. The 'surging' usually associated with flash-steam hydroplanes is now quite absent. It is thought that this is probably due to the pump not having to pump aerated water—not to mention the grit and other solids.

By the way, the small tank marked 'butane fuel' which can be seen overhanging the stern of *T.N.T.* does not contain butane but paraffin/-petrol fuel (see *Plate 25*). In 1972 a new speed record for steam was established at 69 *mph*.

The 'Pilliner' hydroplanes

Bernard Pilliner commenced building flash-steam hydroplanes in 1936, and his brilliant experimental work has made an immense contribution to the subject. His passing in 1970 was indeed a very sad loss to all concerned with model power boats.

The early attempts were not very successful but served as a foundation for further experiments. At first, poppet valve engines were favoured but a change to piston valves showed promise and the most successful engine had a $\frac{3}{8}$-inch diameter overhead piston valve with the 'quick return' valve gear, 1-inch bore and stroke (*Fig. 12.5*). This engine powered four different boats. *Frisky*, 1946; *Ginger*, 1947; *Frolic*, 1951; and *Eega Beeva* 1954. Piston valve stroke was originally $\frac{3}{8}$ of an inch but the special valve gear enabled this to be reduced to $\frac{1}{4}$ of an inch without undue port restriction. The main reason for this change, however, was the heavy loading of the valve gear, and in the final form it was considered safe up to about 9,000 *rpm*. Originally, the cylinder was of 3 per cent nickel steel with uniflow ports, but the nickel was later replaced by cast iron without the ports since no noticeable difference occurred when the uniflow ports were covered up. Incidentally, the wearing qualities of the cast iron proved considerably better than the nickel steel—working in combination with a piston of cast iron. Study of the engine shows substantial proportions without being excessively heavy; built on *ic* engine lines with enclosed crankcase, ball bearings on mains, big end and follower crank. The gear reduction to feed pumps was only $2:8:1$ and this means the pumps were operating at over 3,000 strokes per minute at full speed. Improved

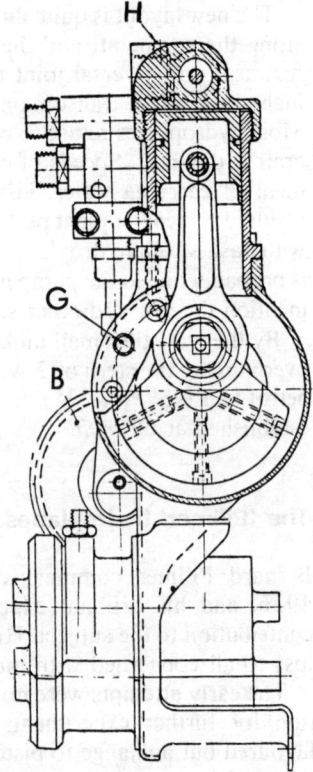

Fig. 12.5 Single-cylinder engine with overhead valve and 'quick return' motion by B. Pilliner. The diagram shows part sectional elevations of *Frolic*. Key: A = Pump drive pinion, B = pump drive gear, C = copper gasket, D = split collet, E = water pump, F = fuel pump jacket, G = oil feed to crankcase, H = oil feed to steam, J = steam annulus, K = exhaust annulus. Note; pumps shown parallel to cylinder are actually at a small angle to make clearance for piping.

valve performances given by the valve gear are discussed in Chapter 6, together with the timing diagrams.

A somewhat daring experiment was tried out in *Frolic*, the engine was sited right aft with the flywheel projecting beyond the transom and having propeller blades projecting from its rim! This worked very well indeed eliminating propeller shaft, universal joints, skeg, etc. Overall diameter over the blades was $6\frac{1}{4}$ inches, on one propeller with the flywheel 'boss' $2\frac{3}{4}$ inches diameter. Although multiple blades were tried out best results were obtained with two blades only.

Large pitches—some over 12 inches—were easily obtained with a very favourable angle of attack. The distance between water level to shaft centre when running was about 2 inches. Once running this seemed to make no difference, but a strong sideways thrust occurred on the getaway. An interesting attempt to prevent this was by offsetting the engine to the near side instead of the central position. This was not

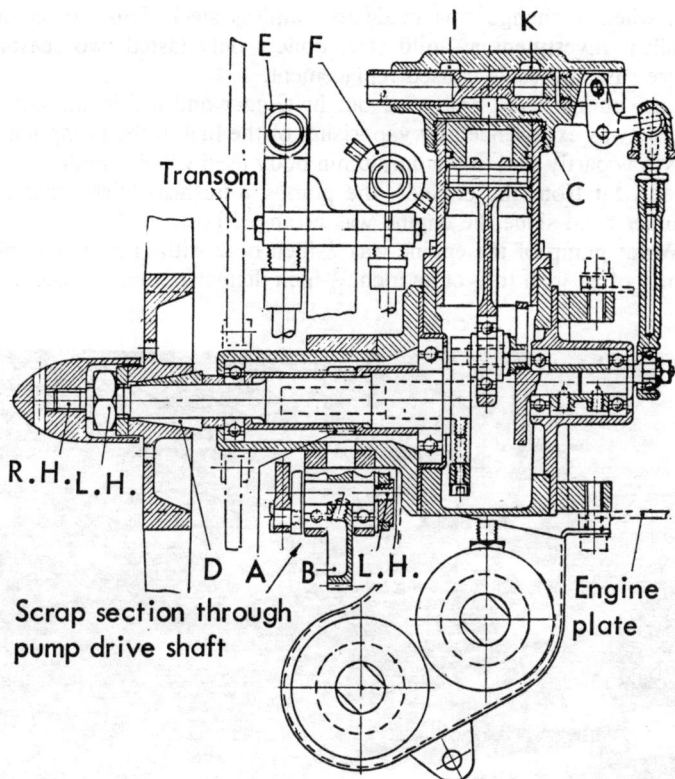

Scrap section through
pump drive shaft

Engine
plate

too successful, and reversion was made to a central propeller in the
next boat.

Some modification was made on *Eega Beeva*. Instead of the flywheel projecting outside the transom, the engine was sited a little
further forward and a normal type of propeller was driven by a short
shaft. This had the effect of bringing the centre of gravity to a more
suitable position, also reducing stresses in the crankshaft.

Some details of the boiler experiments have been given in an earlier
chapter but a few additional comments may be of interest.

The hydroplane *Ginger* had no less than 60 feet of steel tubing in the
boiler, which must have operated very inefficiently compared with the
later craft which had little more than half this amount—33 feet. When
the 60-foot boiler was reduced to 42 feet, an improvement in performance occurred—presumably due to better combustion.

Mild steel was used for all boilers except that for *Eega Beeva* (*Plate*

32), when a change was made to stainless steel. This proved an excellent investment as mild steel boilers only lasted two seasons before rust and scale caused replacement.

Fuel pumps were first tried out in *Ginger* and a fair amount of trouble was experienced by vaporising of the fuel in the pump itself. This was partly due to the duralumin body used which rapidly transferred heat from the engine—the pumps were also fairly near the cylinder head since the engine was arranged horizontally.

Water pump of the engine was $\frac{5}{16}$-inch bore with choice of stroke from $\frac{3}{8}$ of an inch to $\frac{5}{8}$ of an inch, $\frac{3}{16}$-inch diameter stainless steel ball

Plate 35   *Eega Beeva* by B. Pilliner. Note the array of flash steamers in the background

valves restricted to ·015 inch lift. Fuel pump $\frac{1}{16}$-inch bore × 0·2-inch stroke, $\frac{5}{32}$-inch diameter inlet and $\frac{1}{8}$-inch diameter delivery, both valves restricted to ·012 in lift. No oil pump was used, but oil was fed to the crankcase by centrifugal force from one oil tank, while valve and cylinder lubrication was by 'hydrostatic' lubricator operating by water pressure from the cold end of the boiler.

The 'Bamford' hydroplanes

Jim Bamford commenced operations in 1950 with a most interesting and unusual project—a flash-steam hydroplane driven by a turbine. Now very few turbine flash plants had ever been constructed—one of these being the small turbine plant by Prof. D. H. Chaddock referred to at the end of Chapter 6. As far as is known no one had previously done well enough to get round the course at any speed—let alone compete against 30 *cc* petrol engined jobs.

Plate 36   J. Bamford's *Bebug* on simple water brake set-up (this is not the turbine-driven job referred to in the text). Note the rev-counter and stop-watch on right hand side.

The first turbine was of the de Lavel pattern, having a wheel $3\frac{1}{4}$-inch diameter running in small ball races. The reduction gearing to the power take off and pump drive also employed ball races, the whole design being of neat but robust construction. A great deal of bench testing was done on this turbine, first with a 'pot' boiler then with a flash boiler consisting of 20 feet of $\frac{1}{4}$-inch diameter tubing arranged in two parallel coils.

In order to find the B.H.P. developed, a brake of some sort was required—and here emerged one of those strokes of genius. A 2-lb.

syrup tin had two bushes placed centrally, a propeller fixed to a shaft running through the middle, pivoted on a bracket—result—a water brake that worked (*Plate 36*).

After many trials and tribulations 0·47 B.H.P. was developed at 60,000 *rpm* with water evaporation at 9 ozs. per minute. This was not thought to be powerful enough for the job required and the turbine was then retired to the shelf.

It was then decided to build a piston engine in order to get the boat on the water, but the engine chosen was far from conventional. In this design a disc valve operated by a peg on the piston crown admits the steam, exhaust being a straight uniflow through ports in the cylinder wall (see Page 85).

Early trials were conducted on the same water-brake and gave 0·5 B.H.P. at 8,000 *rpm*. Both the piston peg and the disc valve soon showed signs of distress. Finally, a piston of nickel steel and a disc valve made from a motor-cycle exhaust valve proved satisfactory.

The dramatic increase in power given by altering the compression ratio has already been mentioned. The engine was then fitted into *Hero* (1953), using an 'engine forward' layout and a boiler of similar pattern to that used in Bernard Pilliner's *Frolic* at this time—30 feet of $\frac{1}{4}$-inch steel tube in four coils firing by four burners. Piston seizure was in evidence on early trials, and eventually no less than ·007-inch clearance was required before the trouble was cured. An intriguing feature was that a cord was used for starting up—as the engine would run equally well in either direction.

Maximum lap speed was around 40 *mph* with an average of about 30 *mph* for the five laps but the end of each run was marked by the boat sinking—due mainly to lack of freeboard and forward buoyancy.

Since the boiler was capable of evaporating over $1\frac{1}{2}$ lbs. of water per minute it was decided to try the turbine again, but using a type of wheel known as the 'Stumpf' pattern. The buckets for this design were produced by end milling. Although pond tests were not particularly encouraging, the turbine showed some urge and further bench testing was carried out on the complete boat, using the ingenious wind tunnel described in *M.E.*, December 31, 1953. This resulted in a 'blow up' when the turbine casing fractured due to the load of the water pump.

It was now decided to build a new turbine of the 'Stumpf' wheel type (*Fig. 12.6*), using a 3-inch diameter wheel. A speed of 95,000 *rpm* was anticipated and was attained. The new wheel was machined from nitralloy and weighed 14 ozs. First trials were promising, about 2 laps being covered at 45 *mph* before slowing down occurred.

It was then that a new hull and new boiler were constructed. The boiler has been described earlier and in its original form was lagged with asbestos cloth and aluminium foil. In practice running 40 *mph*

was attained on a number of occasions. At the 1954 St. Albans regatta
a time was recorded for the 5 laps—about 32 *mph* average.

As far as is known no other person has achieved anything like these
speeds using a turbine to drive a hydroplane. No praise can be too high
for the tremendous amount of ingenuity and experimental work
required to reach this performance.

Jim Bamford told us recently that he considers experimental turbines designed for high power are rather dangerous as the wheel runs at

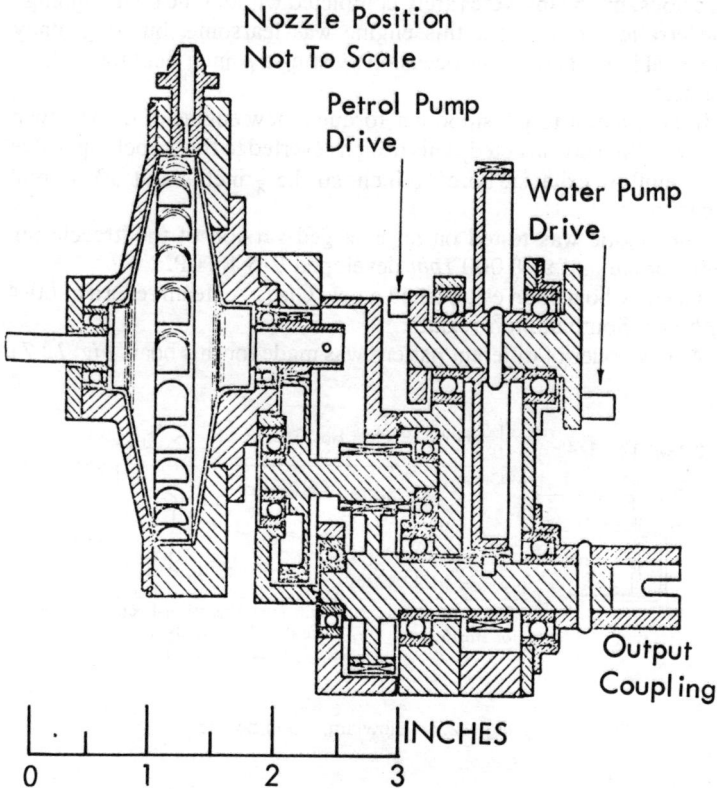

Fig. 12.6   Section of turbine and gear box

its maximum safe speed *under load*. The speed could double off load—
for example, if a reduction gear strips. We think that this warning
should be given, although exponents capable of emulating his work
would probably have prior knowledge of such things!

The main trouble with the turbine was the inability to keep up speed for more than three laps, and it was then decided to make a simple piston valve engine for comparison using the same boiler and hull. Cut-off on this engine was at first 75 per cent stroke. Subsequent reductions brought successive increases in power until the timing period was only 58 degrees total, with cut-off at 52 degrees after T.D.C. This short admission period was responsible for the boiler operating at very high pressures—3,000 *psi* being recorded on one test. Performance was quite sensational as lap speeds of 70 *mph* were recorded on many occasions, but 5 laps were rarely completed without the boat 'flipping'. The erratic torque from this engine was fearsome, bursting many universal joints and on one occasion twisting a $\frac{5}{16}$-inch shaft through 90 degrees!

In an attempt to get smoother torque a new engine of the vee-twin type was then constructed. This design reverted to the 'knock-up' value gear, uniflow exhaust, bore $1\frac{1}{8}$ inch, stroke $\frac{7}{8}$ inch about 30 *cc* total capacity.

The engine was tested on an enlarged version of the 'treacle tin' water brake and at 8,000 *rpm* developed 2·5 B.H.P.

The new boat soon established a new Class-A steam record at 60·4 *mph* at a Southampton regatta.

A new boiler of different pattern was made for this boat (*Fig. 12.7*).

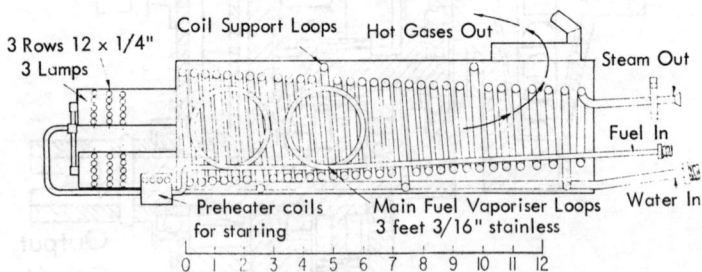

Fig. 12.7   Blowlamps and boilers

It will be noticed that instead of the venturi arrangement three flame tubes blow into an enclosed casing, so that all air passes through the burners. The efficiency of this boiler is not as good but still capable of evaporating well over 1 lb. of water per minute. The main reason for the change was the fact that this design is shorter in length and less susceptible to position, as firing may be from aft without affecting the combustion.

Fuel used is 5 parts of paraffin to 1 of petrol—the latter added to assist starting up. Fuel pressures of over 200 *psi* have been recorded pumping this mixture. This is very much higher than pressures obtained with a bicycle pump!

A more recent engine was constructed in 1970 and represented an attempt to develop the power at a higher speed. Inlet valve was again the piston-operated flap valve, but an additional piston valve provided auxiliary exhaust and prevented excessive compression. The first brake test gave over 2 B.H.P., but the revs would not rise above 8,000 *rpm*. This speed limitation was not known when the engine was designed and seems to represent the limit of the piston-operated valve.

Plate 37   The latest engine by J. Bamford. Two piston valves are driven via a separate shaft geared to the crank-shaft

A new head was then constructed having two piston valves and in its new form developed 1·1 B.H.P. at 7,700 *rpm*. However, it was now found that the water brake tended to cavitate badly at much over 8,000—and the new engine could rev much higher than this. So far the pond tests have not been particularly good but that is all part of the game. This engine is illustrated in *Plate 37*.

The details of the various plants given in this chapter are necessarily short, and there is much information that has had to be omitted. Perusal of these accounts will reveal a good deal of experimental work. In fact, all flash-steam hydroplanes are experimental. Let us quote Jim

Bamford's article in the *M.E.* (July, 2, 1971): "The engine was fitted with the pump gear from the twin, and using the same engine bearers dropped straight into the hull and engaged with the prop shaft perfectly. There is a great deal to be said for standardising on pumps, crankcase sizes and shaft couplings when it comes to experimental work".

It is our earnest hope that this little book will stimulate readers to "go thou and do likewise" and we would like to put on record how much pleasure we ourselves have obtained in our own attempts at *Experimental Flash Steam*.

APPENDIX

Suppliers

T. W. Metals Ltd, Majestic Road, Nursling Estate, Southampton, Hants SO16 0AF. Telephone 023 80 739333 are large scale stockists of stainless steel tube from $\frac{1}{32}$" to $2\frac{1}{2}$" diameter and pipe in much larger sizes. Model engineering suppliers such as Blackgates Engineering, telephone 0113 285 3652 also stock some sizes.

Blowlamp fuel referred to in this book, although available occasionally, has largely been replaced – for flash steam purposes – by paraffin.

References

A large number of articles about flash steam have appeared in the past 60 years in the columns of *Model Engineer*. While few readers are likely to possess sets of volumes personally, some Model Engineering Societies have complete sets, and it is also possible to obtain individual volumes from public libraries by special request.

It was thought that some of these references might be useful. The period covered is approximately from 1923 to 1960.

Steam cars and General applications

1924, Vol. 50, Jan. 10: *A Boiler Control System for Steam Cars.* 1927, Vol. 57, Aug. 11 and 18: *Semi-flash Stationary Plant,* R. H. Bolsolver. 1930, Vol. 62, Feb. 6: *Technical Data—Doble Car.* 1932, Vol. 66, April 7: *Steam Car Matters,* R. H. Bolsolver. 1932, Vol. 66, May 26: *Specification for Steam Car,* A. W. Lambert. 1932, Vol. 66, June 16: *A Light Steam Car,* W. S. Paris. 1935, Vol. 73, Nov. 21: *Temperature Control,* E. T. Westbury. 1937, Vol. 76, March 11: *The Modern Steam Car,* by "A Steam-car Owner". 1937, Vol. 76, April 15: *Besler Rail Car,* Chas. S. Lake (M.I.Mech.E). 1937, Vol. 77, Aug. 12: *Further Notes on Steam Cars,* by "A Steam-Car Owner". 1940,

Vol. 83, Aug. 1, 8, 15 and 22: *Turbine Steam-Car Plant*, A. C. McLeod (A.C.G.I.). 1944, Vol. 91, Nov. 30: *Bolsolver-Rodgers Tractor Conversion*. 1954, Vol. 110, May 6: *A vee-compound Steam Car Engine*, Thos. Hindle. 1955, Vol. 113, Sept. 8: *The Stanley Saga*, George W. McArd. 1956, Vol. 115, Dec. 27: *Modern Steam Cars*, George W. McArd. 1960, Vol. 122, April 21 and May 5: *Is the Steam Bicycle Feasible?* H. E. Rendall.

Model Applications

1923, Vol. 48, April 19: *Power Plants for Model Aeroplanes*, A. F. Houlberg, (A.M.I.Ae.E.). 1924, Vol. 50, May 22: *Sunny Jim III*, A. Norman Thompson. 1925, Vol. 52, May 14: *'Chatterbox' Steam Hydroplane*, S. Clifford. 1931, Vol. 65, July 2: *Concerning Flash Boilers*, E. T. 1933, Vol. 69, July 13: *Model 'Sentinel' Railcar*, L.B.S.C. 1935, Vol. 72, March 7, April 11 and 18, May 23 and 30, 1935, Vol. 73, Sept. 26, Oct. 3 and 10, Nov. 21, 1936, Vol. 74, Jan. 30th, May 7 and 14, a series of articles: *Suggestions for Improving Flash Steam Plants*, E. T. Westbury. 1936, Vol. 74, April 9: *Flash Steam Plant for Model Aeroplanes*, H. H. Groves. 1936, Vol. 75, July 9 and 16, Oct. 15 and 22, November 26, Dec. 3, a series of articles: *Improving Flash Steam Plants*, E. T. Westbury. 1936, Vol. 75, Aug. 20: *Model Destroyer 'Amazon'*, W.F.W. 1938, Vol. 78, May 5: *Engine for 'Tornado IV'*, E. T. Westbury. 1938, Vol. 79, July 28: *H. H. Groves's Engines*, E. T. Westbury. 1940, Vol. 82, March 28: *Coal-Fired Flash Steam Loco*, N. Dewhirst. 1943, Vol. 89, July 8, 22 and 29 August 25: *'Ifit' Series*, A. W. Cockman. 1950, Vol. 103, Dec. 21, 1951, Vol. 104, Jan. 4, 11, 18 and 25, a series of articles: *Experimental Steam Turbine Plant*, D. H. Chaddock. 1951, Vol. 105, Oct 11 and 18: *Steam Turbine Plant*, J. A. Bamford. 1953, Vol. 109, July 23 and 30, Aug 6 and 20, a series: *Flash Steamers*, B. J. Pilliner. 1954, Vol. 110, May 6: *Simple Valve Gear for Flash Steamers*, J. A. Bamford.

Index

Foster, Gerald 40
Frisky 173
Frolic 173, 174, 178
Fuel containers 155
Fuel pressures 181
Fuel pumps 68, 176

Ginger 59, 173, 176
'Goldsworthy Gurney' boiler 4, 5
Groves, H. H. 51, 52, 53, 64, 86, 94, 157, 158

Heating surface 2
Heat treatment 157
Hero 65, 178
High pressure joints 152
High pressures 4, 8, 11, 25
Hindle, Thos. 60, 61, 89, 90, 158
Hobbs, Edward 92
Hodsdon, Alec 15, 42
Home-built steam cars 31*ff.*
Hydraulic pressures 29
Hydroplanes—see *Model hydroplanes*
Hydrostatic air release 58
Hydrostatic lubricator 75, 104

Ifit series 64, 67, 68, 166*ff.*, 170
Incomplete combustion 67
Incubus 44, 156
Indicator diagrams 76
Institution of Mechanical Engineers 25
Irene III 157, 158, 159

Jutton, Frank 62, 64, 68, 82, 171

Keen, Charles F. 38, 39
Kendrew, H. W. 40, 42

Lagging 65
Lake. Chas. S. 25
'L.B.S.C.' 49
Lead, valve 76
Light Steam Power 43
Light Steam Power Society, The British 43
Locomotives, model 48*ff.*, 51
Locomotives, steam 10
Lowne, F. 86, 87
Lubrication 51, 75, 103, 162

Marine boilers, 'Scotch' 2
Martin, A. 160

Materials 55*ff.*, 73, 100, 146
hulls 121, 122
suppliers 183
McArd, Geo. W. 38, 40
Meehanite 74
Metering 97, 105
Mild steel tubing 65
Model aircraft 51*ff.*
Model Boat Power Association 45, 65, 87
Model boats 44*ff.*, 53, 54, 56, 57, 58, 113
Amazon 46
Silver Jubilee 45
Verulam 46, 56, 57, 113
Model cars 51
Model flash boilers 55*ff.*
Model hydroplanes 44*ff.*, 53, 54, 58, 62, 64*ff.*, 66, 68, 115*ff.*
Bamford hydroplanes 177
Bullrush 157
Chatterbox III 44, 159, 170
Eega Beeva 63, 64, 175, 176
Evil Spirit 158
Folly 44, 56
Frisky 173
Frolic 173, 174, 178
Ginger 59, 173, 176
Ifit series 64, 67, 68, 166*ff.*, 170
Ifit 7 64, 170
Ifit 9 67
Incubus 44, 156
Irene III 157, 158, 159
Mystery 158
Pilliner hydroplanes 173
Sunbeam 164*ff.*
Sunny Jim 159
three-point hydroplanes 124
Tornado series 160*ff.*
T.N.T. 172
Vesta 171
Vesta II 171
Model locomotives 48*ff.*, 51
flash boiler 48
pressure tank 49
water feed arrangement 49
Model railcar 49
Model traction engines 51
Model turbines 92
Multi-burner blowlamps 67
Multi-cylinder engines 90
Myer, K. L. 48
Mystery 158